1 Break from

It is easy to get to the top after you get through the
crowd at the bottom. —Zig Ziglar

Why This Book?

If you are a builder or builder's representative, you have an uphill road with many obstacles when it comes to working with brokers. Please understand: there is a lot of anxiety and suspicion of builders among the brokers. Many builders have burned their bridges with the brokers and the damage will take years to repair.

It is not hopeless though, and this is actually good news, if your desire is to contrast your company, and break from the pack of builders who don't understand the power of effectively working with brokers. It will take effort on your part, and the suggestions in this book, although simple, are not all easy. But I can tell you that it will be worth the investment. If you can give a little effort in the area of broker relations you will be greatly rewarded by the brokers' support.

Little Things

Sometimes it's the little things in business that create success or failure. Most automobiles are not taken off the road because of a car accident. They are destroyed or crippled after years of slowly decaying from the acid building up in the motor oil or the natural elements beating down on them day in and day out. Then one day the motor doesn't start. What happened? Lack of preventative maintenance has caught up. In the same way that a car can break down from lack of maintenance, the relationships with your local brokers can be strained or neglected to the point that one day you wake up and there is little or no support from the local real estate community.

When that happens, you will be at a competitive disadvantage to the building companies who are effectively and actively leveraging their local brokers. I have seen some home builders who, when they recognize that their business is failing because of lack of broker support, throw a big broker party or offer huge broker bonuses. This, although well intentioned, is equivalent to the man who neglects his wife or girlfriend for years, and then, just before she leaves him, buys a dozen roses, a box of chocolates, and a huge stuffed animal.

The result is usually suspicion and disbelief. The ideas in this book are designed to be effective and difficult to reproduce, which will give your company the competitive advantage. In other words, the idea is to outsmart, not out spend, your competition. The separation in this business is in the preparation. In most any market it is the little things that can make a big difference and create a tipping point in your favor. As the saying goes, "look to the littles." This book will give you real, hard won, practical suggestions to assist and prepare your company to create long-term, powerful, and productive relationships with your local real estate brokers.

Why Now?

When the real estate market was collapsing in 2007, I had no Plan B. In other words, if I couldn't make it in real estate, I was quite sure that there were no other jobs in my area that could support the personal and professional expenses that I had created during the good times. I was the proverbial grasshopper, and I could feel the cool winter breeze on the horizon. I did, however, have several things going for me. First, I had become a specialist in new home sales. This was important because builder representatives were leaving faster than they could be replaced. Second, I had acquired the skills to effectively work with the local brokers to get qualified buyers.

This was the single factor that allowed me, and the builder I represented, to not only survive during the hard times, but to thrive. As the market starts to warm up, I believe many builders will get sloppy and apathetic about the principles of sales and marketing success, such as leveraging the power and influence of the brokers. Now is the time to master these principles to be prepared to weather the inevitable market changes.

If It Doesn't Work, Drop It

Around the time of the real estate crash in 2007, I interviewed for a sales position for a new home building company. The builder asked me how much advertising I would do if he let me represent his company. (During the housing boom, if you were a licensed broker and you wanted to represent a builder or get a listing, the formula was simple: whoever promised to pay for the most advertising usually got the listing.) I stated that I would not advertise his homes. In fact, my recommendation was that he cancel his advertising all together. This startled the builder, but he was intrigued by my suggestion. (If you are a builder, please do not cancel your advertising. I am using this example from my past to simply make a memorable point. Sometimes we get into an advertising and marketing rut, and when times get really tough, as they did in 2007, drastic measures are warranted.) With the market collapse, everything changed. If something didn't work, it had to be cut.

I am not against advertising. In fact, I know that, if used correctly, it can be a powerful tool. I do have some observations, though. First, it can be easily copied. If company A has a billboard, company B can get two billboards. If company A has a full-page ad, company B can close the gap and simply pay for two full-page ads. Second, advertising is expensive. Many builders don't realize what you can do for the price of a billboard when it comes to guerrilla marketing. We see some of the national builders unleashing a glitzy advertising campaign and it is tempting just to fall in line and copy. Remember, David killed Goliath not because he copied him, but because he used tactics that were unconventional and bewildering to his competitor.

David was a specialist, and his skill was not easily copied. In fact, the giant probably never saw the rock that hit him. If you have a great advertising campaign and it can be copied, what advantage is that? The ideas and suggestions in this book, just like David's smooth rocks in a leather sling, may not seem powerful and exciting at first glance, but they have the power to give you the edge. I am writing this book now as a sort of handbook that builders can use to create alliances with the broker community. Think of your relationships with the brokers as a long-term strategic marketing plan.

Take Away:

- If it doesn't work, drop it.
- Stay sharp. Don't let the market increase allow you to get sloppy.
- Be efficient. Utilize the most cost-effective traffic building methods.
- Make it your goal to outsmart rather than outspend the competition.

Why Builders Need Brokers

You pay the broker when the home sells. You pay the bank when it doesn't. —*Tom Richey*

Why Do Builders Need Broker Support?

The answer is simple: the brokers have the buyers. With more than one million Realtors active in the United States, many builders are finding it difficult to sell their homes without the brokers' support. If you were in search of pearls, you would look for oysters, right? The pearls are in the oysters. Let me say it again: the brokers have the buyers.

Do some market research in your area. Find out how many new homes were sold last year versus used houses. Almost without fail in every market, the number of used house sales dwarfs the new home sales. Is that because there are more used houses on the market? That may be a reason, but I believe there are other factors at work. I'm sure if you did a survey more new beds were sold nationally than used beds were last year. I am sure more new shoes were sold. Is it because used houses are a much better deal? I say no! In many markets, the overdeveloped land has gone down in price and builders are able to be competitive price-wise with used homes. Nearly every buyer would rather have a new product over a used one, so why are more used houses selling? I believe it's because the buyers in many instances never even get a chance to see the new homes. The brokers have the buyers, and, for a variety of reasons, they are not even giving your company a chance to talk to their buyers. Remember this point: your biggest competition is used houses!

Reasons Brokers Sell Used Homes

The brokers have the buyers, and they are selling them used houses.

Here are some of the reasons:

1. Brokers can make a quick sale and get paid faster selling an existing used house than waiting for a new home to be built.
2. Many times they can just show the homes listed in the MLS.
3. In many cases, brokers simply don't know where your homes are or that you even exist.
4. Brokers fear losing control of the sale.
5. Brokers are uncomfortable practicing outside of their area of expertise.

Take Away:

- The brokers have the buyers.

- Your biggest competition is used houses.

- Engage a marketing strategy to lock hands, not horns, with brokers

3 Where Brokers Win

When everything seems to be going against you, remember that the airplane takes off against the wind, not with it. —**Henry Ford**

Why?

So why and how do the brokers have the vast majority of buyers? Consider the following areas where brokers dominate builders.

Brokers Dominate Relocation

Think about who coordinates with most relocation companies. It's the brokers. USAA is a good example. Many military families use USAA for insurance and financing. They have key relocation real estate brokers whom they refer their customers to. To my knowledge, they are not working with any local or national builders when it comes to referring their customers. Better get a couple more high-gloss ads if you want to compete with USAA. The bottom line is that brokers dominate relocations and they will continue to do so for the foreseeable future.

Even buyers not working through a relocation company will choose to work with brokers. Most buyers, when moving to a new city, simply call one of the top brokerages in that area. Many buyers prefer to work with a broker because the broker usually knows the area and schools and can take them on a tour and show them around. How many times have you picked up a potential customer from the airport? I know of brokers who offer these kinds of services, and these brokers will capture many qualified buyers. The question is: will these brokers take their buyers to see your homes?

Brokers Win on Search Engines

One of the most popular internet searches prospective buyers use when looking for homes is the name of the town, along with the words "real estate." For instance, the most searched term for people looking for homes in Phoenix would be "Phoenix real estate." When I type in almost any city with the words with "real estate" or "homes" after it, without exception the top results on the major search engines are Zillow, Trulia, Realtor. com or a large brokerage such as Keller Williams or RE/MAX. I do find builders, usually large ones, in the top part for paid advertisers. However, most people usually skip over these advertisements and click the ones that come up organically, and those are almost always the brokers. If you click into Zillow, Realtor.com, or Trulia a list of smiling brokers appears, seemingly ready to answer any of your questions.

Brokers Have Superior Signage

How many resale homes are for sale in your market? The answer will correspond closely with the number of real estate signs. That is because each of the properties active on the market will have a real estate sale sign in the front yard with a broker's number. Besides brokers having many more signs, most buyers will be more inclined to call from a broker's used listing than a new home sign anyway. Take a look around most new home communities. You will probably see dust, trash, earthmovers, dumpsters, and workers. Contrast that with the large cherry tree in front of the used house with a brokerage sign beckoning the buyers to call and learn more about the home. Bottom line, in nearly any given market, the brokers have you beat when it comes to signage. The result is that the builders are missing out on these buyers, if they are not coordinating with the brokers.

Brokers Capture the Buyers with Homes to Sell

When I talk with new home sales people, one of the biggest challenges they face is that many of their buyers have a home to sell. Most of these buyers are not trying to sell their home "for sale by owner," so they are listed with a broker. According to the National Association of Realtors, 89 percent of homes sold in 2015 were sold using an agent or broker.

Understand that when the home sells, the broker will now be in a position of influence and power because they just helped their customers navigate the troubled waters of selling their home. The brokers are the first to capture the leads that come from the sale of the used houses in your area.

Brokers Usually Have Strong Relationships with Their Buyers

Builder's Broker Policy: If your broker is not with you on the first visit, he or she will not be paid on the transaction.

Translation: You can go tell your Aunt Gina, "Tough Cookies!" The first thing a real estate broker does when he or she gets his or her license is to contact his or her friends and family members to let them know he or she is a real estate broker. Real estate, for most people, is intimidating and stressful. According to experts, the third most stressful thing a person can do is move. Some believe that "home" is the second most emotional word in the English dictionary, just behind "mom." The bottom line is that for the large majority of the population, buying and selling real estate is stressful and emotional. If they have a family member who is in the business, you better believe they will contact that person for advice. Your Builder-Broker Policy is not stronger than these family relationships. Brokers have a sphere of influence; and it is called a sphere of influence because they can influence the buying decisions.

Ability to Show Used

Many buyers can't completely see the advantages of a new home until they can see a number of used houses. Can your sales staff take a few weekends off to show used houses to the buyers who are on the fence so they can make a confident buying decision? Of course not! Brokers have the ability to show the used houses and bring the buyers to you for a quick and decisive sale. The average broker is working with 3 to 5 buyers at any given time. The number of brokers in any given market dwarfs the number of new home sales people. Brokers have the advantage of a much larger pool of potential buyers.

Social Media

At almost every builders' conference I go to, social media is being touted as "The Thing" to save the industry. There are many successes when it comes to social media, but on the whole, if many builders were honest, the results have been lackluster considering the emphasis given to this new medium. Many marketers will say, "That is because you are not doing it enough, or you are not doing it right." Many will encourage us to double down on our social media efforts. I am very much in favor of social media. I do, however, want to suggest that there may be another factor at work here. It is the fact that the brokers have us beat, and beat big, when it comes to social media. Here are just a few reasons why:

1. **Percentage.** According to the National Association of Realtors, 91 percent of Realtors are actively leveraging social media. In contrast, according to Professional Builder magazine 45 percent of builders are utilizing social media. It goes on to say, "Despite the rapid growth, social media still plays a relatively minor role in generating new leads for home builders. Just 6.9 percent of builders said that a significant portion (26 percent or more) of their new business is coming from social media efforts. So the percentage of builders effectively using this medium is much lower than that of the brokers.
2. **Numbers.** There are many more brokers in any given city than there are builder's representatives. In the city where I live, it is about a 50 to 1 advantage for the brokers. In other words, there are 50 brokers for every 1 new home sales representative.
3. **Quality.** Brokers as a whole are just plain better at social media. Most builders in my experience are left-brained and more inclined toward the science of building and profit

creation rather than the nuances of social media. You say, "Quint, that's why we have hired a company to do our social media for us." The problem is that most people can see right through these social media blitz campaigns, and often those efforts can create a disingenuous feeling. There is a new push in the industry to have the new home sales staff take charge of the social media. If you have been in the business for more than a few years you know that there is turnover when it comes to new homes sales.Great, so now your sales rep has the control over your social media, and he or she is working for your competitor. Just today I was talking with a sales rep from another builder and he is in control of that builder's social media. I asked him, "What are you going to do if you leave your builder?" The response? "I will just keep control over the account."

I am just highlighting some of the pitfalls that I have seen in real life. Again, I am very much in favor of utilizing social media. Obviously this is a quickly-evolving facet of our business that I expect will become more and more a part of a successful marketing plan. I just want to suggest that if you are a builder who hasn't mastered social media, let not your heart be troubled. There are ways to compete with the building companies that have hip and slick campaigns, namely, by leveraging the experts in social media: the brokers.

Community Outreach

Go to the local farmers' market and you will probably see local real estate brokers talking to the local public about the real estate market and handing out business cards. Go to any social gathering and you will see local real estate brokers with their name badges on, socializing and making friends. Check out the Chamber of Commerce, the Kiwanis club, the Toastmasters, and any and all networking groups, and you will find brokers actively working away at making contacts. This is also true when it comes to charitable organizations such as the March of Dimes and "Tough Enough to Wear Pink" campaign. Brokers are usually on the front lines promoting their favorite charity and letting people know that if, "You or anyone you know is in the market for a home, I would be happy to help." Yes, builders have community service programs and do a great deal of good in the community, but the brokers usually have us beaten when it comes to actually getting contacts and buyers from these endeavors.

Office Locations

Drive through a city and try to find the main offices of home builders. They are usually intentionally hard to find because they don't want people coming by interrupting business. The sales are usually done in the actual communities where their homes are being built. Furthermore, these communities are difficult to find, look like a construction zone, and are often not yet updated on Google maps. Contrast that with your local real estate companies. They are always competing for the most expensive, most visible, and easiest to find prime commercial real estate locations. They will in most situations capture the buyers first, which is why you want them on your side.

Take Away:

- The brokers will capture an overwhelming number of the buyers in any given market
- Because the brokers dominate in many areas of capturing new business, it is essential that we get them on our side.
- In today's high-cost, high-stakes marketing, broker cooperation is a safer bet, because they don't get paid until the home closes.

4Synergy!

$$2 + 2 = 5$$

Synergy is defined as, "the interaction or cooperation of two or more organizations, substances, or other agents to produce a combined effect greater than the sum of their separate effects." In other words, leverage. The premise behind this book is that you will make more sales and increase market share by leveraging the power of real estate brokers than by trying to go around them or selling without their support.

My Advice

I recently read about a marketing professional who was in disagreement about the focus on working with brokers. His claim was that if builders weren't paying the brokers, they could pay for the biggest and best marketing campaign. His advice: "If your sales people can't sell without brokers, you should fire them." My advice is actually the exact opposite. If your sales person can't or won't leverage the local broker he or she needs to be replaced. Although well intentioned, I believe our marketing friend was forgetting one detail: brokers don't get paid until a sale is completed, and the deal is funded. No billboard company or magazines that I am aware of will work under the agreement that they will only get paid if their advertisement actually sells the homes they are advertising.

I am not aware of a website builder or contact relationship manager (CRM) company that will put in a 40-hour work week with no guarantee of pay. Many brokers I know work many hours over the standard 40 hours a week and get nothing, unless the homes they sell fund and record.

An Objection

"But if I do enough advertising and have great signage, won't the buyers just go around their broker and buy directly from me?" Yes, in some cases this will happen, but it is usually the exception rather than the rule. You may think that I have a defeatist attitude. Quite the opposite! I encourage self-marketing, websites, social media, CRM, and general outreach to the community. I personally work very hard to self-prospect and find my own buyers and I encourage you to do the same. The fact remains, though, that the brokers now and for the foreseeable future have a stronghold in these areas. For me, it's the difference between walking and taking a cab. Taking the cab is not free, but it is well worth the investment if I can get to my destination more quickly and without being drenched with sweat. When writing this book I am not using a typewriter. Is this because typewriters don't work? No, it's just that using a computer is more efficient. I am sure that most builders are capable of framing, painting, and landscaping their own homes. There are many builders who do that. Most, however, enjoy delegating these tasks to contractors who specialize in that trade. It is no different with the brokers. I see the method described in this book as delegating and creating leverage, rather than surrendering. Most builders are more effective and profitable focusing on land acquisition and product development, rather than networking at the local farmers market and engaging in social media.

Take Away:

- The premise behind *Partnering with Brokers to Win More Sales* is that you will make more sales by effectively working with real estate brokers than by trying to sell your homes directly to the public without the brokers' support.

- Working with brokers is not about defeat and surrender, but rather delegation and synergy.

5 Why the Brokers Need Us

Teamwork makes the dream work.
—Unknown

Why do brokers need builders? The answer is simple. Builders have the homes. Not only do we have the homes, we have the new homes. The content in this book may give the impression that I am sympathetic and defensive of the real estate broker community and, frankly, I am. However, I am equally defensive and sympathetic toward the builders' new home sales reps who feel beat up by the brokers and are tempted to crumble under the power of their juggernaut. First, brokers need to earn their commission. They need to bring you qualified buyers and the home needs to close and fund before they are paid. You have and build the new homes, and many buyers, given the opportunity, would prefer new over used. That gives you, the builder, a lot of power and leverage.

Point It Out to a Broker

When I make a presentation to the brokers, the first thing I do is outline some of the challenges in the used housing market. I want to accentuate the hassles involved with selling used houses. I might ask, "Please raise your hand if you have had a contract for sale on a used house cancel because of a home inspection." Usually everyone in the room raises his or her hand. Then I say, "Now, raise your hand if you have ever lost a sale because of a home inspection on a brand new home." Invariably, not one hand goes up. That makes a powerful statement in support of the new homes.

Next, I try to show the great opportunity available to the brokers when it comes to new homes. In a real estate market such as El Paso, Texas, you will have approximately 1,000 real estate brokers and there will be 3,000 homes actively on the market. I like to ask, "How many homes is that for each broker if sales were divided equally?" The answer is three. Then I ask, "How many in this room could survive on selling three homes a year?" Usually nobody raises his or her hand. Next I ask, "Do you have any idea how many potential new homes could be built in new home communities over the next 10 years?" Usually nobody in the room will have any idea. You can obtain this information from the local home builders' association or even by calling the key developers in an area. The answer is generally in the thousands or even tens of thousands. That is because, if the sales continue, the developer can simply bring to market and develop the land in waiting. This number is usually much higher than most people imagine. Then I ask, "How many real estate brokers are positioning themselves as new home specialists in this town?" Usually the answer is something like 10–20 people in the small markets or double or triple that in medium markets. "How many homes is that per New Home Specialist?" The answer is something like 1,000 per specialist. This is when I explain that each market is different; this is not an exact science, but rather a simple illustration to make the point that a lot of new homes will be sold over the next few years. The brokers who are positioned to specialize and sell these new homes will be the clear winners. The effect of sharing this perspective is powerful and usually gets the brokers excited to start selling more new homes.

Take Away:

- The used houses are not reproducing themselves, so the brokers need new homes.
- This puts builders in a position of power and influence.

6 What You Will Need

There is one body, but many parts.
—1 Corinthians 12:12

What to Look for In a Representative for New Home Sales

In today's competitive market it is essential that we hire and retain the right people in the right positions. It is beyond the scope of this book to cover how to find, recruit, hire, train, and retain top sales people. There are, however, several key characteristics or qualities that I believe a builder representative should possess in order to implement the suggestions in this book.

Ability to Present

Beyond the obvious (well-groomed, high energy, positive, and good communication skills) a builder representative needs to be able to speak in public and present his or her story and USPs (unique selling points) to the real estate brokers. This will be key to your success when it comes to working with brokers. Many struggle with public speaking, but there are ways that we can practice and improve. Joining a local Toastmasters group, attending a Dale Carnegie seminar, or taking a class on public speaking at the local community college can be a great start.

Sometimes a simple change in perspective can be the catalyst to make someone a great public speaker. What if I offered you or your sales team $1,000 to get up in front of a group of brokers and give your company story and outline the top five value differences of your homes? Would you or they be more willing to do it? For most, the answer is yes! What if I told you that doing a broker presentation, either at your model home or at their brokerage during their sales meeting, was the equivalent in marketing effectiveness to having several bill boards, which can cost thousands of dollars a month? In addition, effectively presenting to the brokers has the potential to dramatically increase your sales, which is worth much more than a mere $1,000. This has been my experience, and yet many builders and their representatives still opt to sit on the sidelines when it comes to getting in front of the brokers. I believe many choose to avoid presenting in front of the brokers because it is just plain hard. But, if you wanted an easy job or profession my guess is that you wouldn't be in the home building or sales business. Since we are in the business of new home sales, let's do the hard things that others won't do. Let's break from the pack!

You Need a Voice

You or your company representative needs to be able to effectively do presentations, both for the buyers who may enter your model home, but also for the brokers' sales meeting. I will cover the ingredients of an effective brokers' presentation later in this book, but the willingness and basic public speaking skills need to be there.

Let's say you love your sales representatives, but they won't have anything to do with public speaking or presenting. It may be wise to keep them and just hire someone who can speak for the broker visits. Bottom line: your company needs an ambassador or representative who can help get the brokers on your side.

Take Away:

- Doing an office presentation can be the equivalent to thousands of dollars of marketing and advertising.

- Your company needs a representative who can effectively communicate your USPs to the brokers.

Part Two:
Building Rapport

7 Building Relationships: A Top Down Approach

S.A.L.E.S: Service And Leadership Equal Success.
—Quint Lears

The qualifying or managing brokers of a real estate company are like the guard dogs of their brokers. It is important that you know them, but it is even more important that they know you. A simple call and visit is all you need. "My name is _____, I am a new home builder, and we are building in your area. I would like to visit with you to see how I can support your brokers and your company in selling more new homes this year." Very important: don't assume that everyone knows you or your company. They don't. Even if you have been building for 30 years in your market, there will be people and even brokers who don't know your company.

Take the time to explain who you are. It is important to get the qualifying broker (QB) on your side. Then, when you start to implement your broker outreach, he or she feels included and will help to support your efforts. The main thing is that you don't want them working against you. My experience is that QBs are usually high-ego personalities who have the power to shut down your opportunity before you begin. It took me more than five years to get the opportunity to present in front of a large brokerage in my town. My real estate license was with a different brokerage and the QB where I was trying to get the opportunity to present perceived me as a competitor and a threat. It wasn't until he sold the brokerage that I was able to get in and present. Understand that many brokers are brand loyal. Many have drunk the proverbial Kool-Aid of their particular brokerage brand. Personally, I have found it effective to remain as broker neutral as possible. By law I am required to put the brokerage name and number on my card; but it doesn't have to be on the front of the card and it doesn't have to be the biggest thing on my card. When I am speaking with the QB, I make sure he or she knows I am there to talk about my builder and help his or her agent sell more new homes, not talk about the advantages of the particular brokerage brand I am affiliated with. Obviously, if your sales staff is not licensed this will not be an issue. The point is that a qualifying broker has the power to keep you out of his or her office, so you need to get him or her on your side.

Remember the Office Staff

It is also important to get to know the front office staff of a brokerage. When I am dropping off flyers for the brokers, I give the front staff a gift card and thank them for their work and support. The front desk personnel are the gatekeepers to the QB, who again is the guard dog for the brokers. Get him or her on your side! This is key to getting access to valuable presentation time in front of the brokers.

You Are Not the Only One

Also, understand that many people, companies, and organizations are trying to do the same thing you are. This is because brokers are connectors. They are in the center of a lot of business for a lot of people. They have massive influence over their buyers when it comes to choosing a title company, lender, or even a trade professional such as a landscaper. All these and more will be competing for time in front of the brokers. Additionally, charities, local school fundraisers, and even politicians are vying for the attention of the broker. So how do you win? The answer is that you have the product and systems in place that can help them make more money. That means you are a giver and not a taker. The other people and organizations don't give any money to the brokers, but you do. That gives you an advantage and it is important that we capitalize on this.

What Next?

If you get past the gatekeepers and the guard dogs, it is now time to start working effectively with the brokers.

Take Away:

- There is a lot of competition for the attention of the real estate broker community.

- Get the managing brokers on your side.

- Take great care of the office staff—they can be of great assistance to you.

8 Make the Broker Your Customer

Sometimes a new pair of glasses is needed if we want to effectively work with brokers. I believe we should look to the multiplying power and leverage potential of the brokers. *Partnering with Brokers to Win More Sales* views the broker community as trusted partners working to bring qualified buyers to invest in our new homes. When we have this perspective, it is easy to want to get to know and support them.

Know Your Customer

Getting to know your local brokers is one of the easiest, most effective, and free strategies to increase broker support for your company. All it takes is a little time and thoughtfulness. Knowing personal and professional details about your local brokers will give you the edge when it comes to establishing rapport and respect. When other builders don't even care to know the names of the local brokers and you are able to thank them by name for their military service, you will win the support and respect of the broker community.

Once, when doing a broker presentation, I thanked a broker who had two sons serving in the U.S. Marines for her sacrifice. Since I am a prior serviceman, this was a sincere gesture and it was very well received. How did I get this information? I simply went to the brokerage website and read the bios of the brokers before I did my presentation. You may be surprised by the background and education of the agents in your area. This is important for two reasons. First, it fosters authentic respect for them, both personally and professionally. Second, it gives you conversational topics and sets the groundwork for mutual respect and cooperation.

It's Common Sense!

This is common sense when it comes to doing market research to learn about the potential buyers in your area. It's the first thing good marketing and sales professionals should do. So why do we often ignore this with the brokers? I believe it's because many builders don't look at the brokers as their customers. In fact, many builders look at brokers as if they were the competition rather than collaborators. Remember, they are only your competition if they are not on your side!

There was a builder in my area who would tell buyers to come back without their broker and he could discount the price of the home. It may have worked a few times, but the brokers quickly learned the builder's tactic, and then they banded together, directing buyers away from that builder. Trying to cut brokers out of deals is not a long-term, successful approach to selling more new homes. You should make the brokers your customers. Get them on your side.

Foster an Attitude of Service

It is important that we communicate our message to the broker community through our words and our actions. Foster the attitude in your organization that you want to serve the broker community and help them to be more successful when it comes to selling new homes. Brokers as a whole, in my experience, are very interested in new home sales, and if you approach them in a spirit of humility and service, they will want to support your efforts.

Don't Alienate the Brokers

It can be a costly trap to cut the brokers out of a sale where they are representing a buyer who is interested in your homes.

Never underestimate a broker's relationship with your potential home buyer, and therefore his or her ability to help or hinder a potential sale. In many instances the broker just helped the buyer sell his or her existing home. If so, he or she has usually built a tight bond with the buyers. Trying to interfere with his or her relationship is not that much different than trying to convince the buyer not to listen to his or her father, mother, or Uncle Jimmy. If you have the local brokers' support, it is the marketing equivalent to having a golden goose. Cutting the broker out of a deal is like cutting open the goose to get all the golden eggs. As the story goes, the goose dies and no more golden eggs are laid. When the broker support dies, we lose a pipeline of opportunities with qualified buyers.

One local builder put a sign up that said, "We do not work with brokers." This was more than 10 years ago. The new home sale representative who is still with the company said, "Fortunately, many of the brokers from back then are starting to die off or get out of the business." But he admitted that many still remembered this slight to the broker community and it has taken them more than a decade to reverse the damage of this marketing blunder. Interestingly, this same company now offers large broker bonuses. I guess the company wised up to the fact that the brokers have the buyers.

Understand What It Takes to Be a Broker

It is important to understand the trials and tribulations of being a broker. Taking the time to understand what it takes to be a broker and get qualified buyers will build appreciation and respect for the work they do.

To be a licensed broker, there are national fees, local fees, MLS fees, errors and omissions insurance, continuing education, desk fees, administration fees, signs, business cards, advertising, and much more. And that is just to get the doors open and be in business. To get a buyer, he or she does a tremendous amount of work. Phone calls, emails, sending out mailers, picking up out-of-town buyers from the airport, attending networking events ad-infinitum. The good brokers have taken the buyers to a lender to get them prequalified and have learned about their wants and needs. Buyers these days are few and far between, so when brokers do get a buyer, they are very protective. After the brokers show their buyers 10–20 homes, the buyers call a local builder who tells them to come back without their broker and the builder can reduce the price by $5,000. When this happens to a broker, he or she doesn't easily forget. Again, brokers are a tight-knit group of professionals and will quickly tell their co-workers to avoid such a builder.

Trap: Brokers Get Paid for Not Much Work

It is a trap to think that a broker is just coming in and making a quick commission for doing nothing. Understand that even if the broker has showed 10 used homes and then the buyers come to see you, they wouldn't be in a mindset to buy if the broker hadn't showed them and eliminated the used houses. Many buyers can't see the value of new until they see the used houses.

The Work Involved

It is hard for many builders to see the value in cooperating with Realtors because they truly don't understand the behind-the-scenes work. Let me put it this way: I have met plenty of builders who think that the brokers make a lot of money for doing nothing, but I haven't met many builders who are interested in becoming a broker.

Invest the time to get to know and understand the broker community. Beyond just increasing sales, you will invariably make many friends along the way. I have found this to be true and I hope you experience this as well.

Take Away:

- Many buyers can't see the value of new until they see the used houses.

- Seek to truly understand and know the brokers in your community.

- Make the brokers your customers. Get them on your side.

- Trying to cut brokers out of deals is not a long-term, successful approach to selling more new homes.

9 Appreciation and Recognition

People work for money, but go the extra mile for recognition, praise, and rewards. —Dale Carnegie

Brokers are usually great supporters of various charities and community organizations. Why not thank them for their contributions? If you read in a broker's bio that he or she supports the local animal shelter, send a thank you card for the community support. Here is an example letter:

Cynthia,
I just wanted to say thank you for your community involvement in supporting the local animal rescue shelter. Keep up the great work!

—Your signature

I intentionally made this letter short and sweet to show you that it doesn't have to be fancy or long. A simple thank you is enough.

Go Deep

Not going deeper and more personal with your broker community is a mistake. Brokers usually have a high emotional intelligence and can sense if you are just after their buyers, or if you really care. As the saying goes, "People don't care how much you know until they know how much you care." Most brokers are very appreciative of any recognition. If they didn't want people to know about their community support, they wouldn't put it on their website.

Utilizing Competiveness

Brokers can also be fiercely competitive. Why not put a small ad in the paper listing the top brokers for that month? It is a good idea to get their permission and ask them to send you a photo of their choice.

You could feature a broker of the month on your website. Then you tell the QB of the brokerage that one of his or her agents has been selected to be the broker of the month. Of course everyone will have to check out your site to see it. And hey, since they are on your site, why not look at some floor plans and maybe view the available homes list?

Praise in Public

When a broker comes into the model with a buyer, immediately thank him or her by name for coming by. When you are introduced to the buyers, congratulate them for having great representation:

"Mr. and Mrs. Jones,
I just wanted to let you know that you have great representation. Sarah is one of the best and most respected brokers in our city, and she has a great reputation for taking excellent care of her buyers. [This must be sincere, so this is where doing research and knowing your brokers comes in.] *What I appreciate is that she is taking the time to show you the new homes in the area. Many agents just show what's in the MLS. She has taken the time to learn about the new homes and you are getting to view the homes that many buyers don't even get to see."*

The broker usually will smile and say, "I didn't pay him to say that." The buyers are usually very happy to hear you praising their broker and will often pipe in with their own words of affirmation; "Oh, yes we couldn't be happier with Nancy, she has been working her tail off for us. We feel very lucky to have connected with her." I learned this technique from Steven Romano, a fellow new home sales professional and I can tell you it really works!

Praise and Thanks After the Sale

After the new home closes, I recommend sending a thank you/ congratulations card to the broker who represented the buyer during the transaction as well as his or her qualifying broker. It doesn't have to be long and drawn out. A simple thank you is enough. The letter to the QB should also encourage him or her to congratulate his or her broker. Qualifying brokers are very interested in motivating and giving praise to their agents so they will be grateful for you helping in this area.

Sample letter to the buyer's broker:

Dear _____,

I just wanted to send you a card to congratulate you on your recent sale and let you know how much we appreciate you trusting us with your buyers. It is a pleasure to do business with you.

Sincerely, _____

...

Sample letter to the qualifying broker:

Dear _____,

Just wanted to send you a card to let you know that _____ from your office just sold one of our new homes. We appreciate your brokers trusting us with their buyers. We are always impressed with the professionalism of your agents. Please be sure to congratulate at your next meeting!

All the Best!

Little Things

Again, *Partnering with Brokers to Win More Sales* is not meant to be just a list of ideas to follow, but rather a catalyst to stimulate your own mental creative factory to find new ways to connect with and show appreciation for your broker community. Sometimes it's the little things that matter more than the big. For example, I have found cards of encouragement that I have sent to brokers hanging on their walls years after they were sent. I have never found one of my advertisements pinned to the wall. The focus of this book is to generate ideas that have staying power. Too many of our marketing efforts are focused on us and our ego and usually have, as Bill Gates likes to say about intellectual property, "the shelf life of a banana." Let's focus on things that grow and last. In my experience, people and relationships are what fall into that category. I have found that my business is filled with purpose and fulfillment when I am focused on others.

Take Away:

- Brokers appreciate recognition.

- After a sale be sure to thank the QB for his or her support and remind them to congratulate his or her broker at the next sales meeting.

- Make the brokers your customers. Get them on your side.

- Remember it doesn't have to be fancy or expensive. The little things can be big things.

10 Look to the "Littles"

Nobody trips over a mountain. We trip over the pebbles. Learn to walk over the pebbles and you will find that you have covered the mountains.
—Unknown

One of the biggest fears brokers have is losing control of the sale. Why not communicate more often, even if it is about the inconsequential? If a buyer comes by to measure the refrigerator and take a few pictures of the model home, give a quick call or text to his or her broker representative, if he or she has one, and let him or her know. "Linda, just wanted to give you a quick courtesy call and let you know that your buyer, John, came by today and measured for a refrigerator. He decided that he is going to purchase his own refrigerator. You don't need to call back; I just wanted to keep you in the loop and give you a quick update." The next time the broker sees the buyer, he or she can say, "Were you able to find the refrigerator you wanted at the store?" The broker feels in control; the buyer feels supported.

Communicate the Little Things

I once heard a quote that said, "If people like you, they will listen to you; if they trust you, they will do business with you." I believe that in sales there is too much emphasis on being likeable and not enough effort on building trust. Communicating the little things is one way to build trust, and it costs nothing but a little time and consideration. I have had brokers call me when they were on vacation and say, "Quint, I am out of town; can you meet with my people? I trust you." Imagine if you had a good percentage of agents who would send you the contact information and recommend their buyers to contact and talk to you. Would that be a plus for your business? When most builders can't even get the brokers to view their homes, you will have an advantage that is difficult to reproduce.

Making Changes to the Contract Without Asking

Doing addendums that change the purchase price without keeping the broker in the loop is a mistake. At best, it irritates the broker. At worst, the buyers are raising the price past where they are financially capable. The broker is usually in close contact with the lender and will help to guide less experienced buyers who may price themselves out of buying a house away from you. It is better to have their broker pull their spending reins back instead of you. Many new home sales people jeopardize the sales when they are not in close communication with the buyer's broker representative. Understand that the broker wants to keep the deal together as much as you do, and he or she can often guide the sale, and have influence over the buyer when it matters most.

Not taking my own advice, I did an addendum recently with a buyer and I didn't let the buyer's broker know. Not realizing the addendum was already signed, the buyer's broker organized a meeting with the buyer and me to sign the addendum. The buyer let him know that the addendum was already signed and the broker, who is a friend, called me very annoyed. The exchange made him look out of the loop with his buyer. Because I was busy and friends with the buyer's broker, I took for granted the principle of communicating the little things and didn't make the extra effort to keep him in the loop. This is a mistake you don't want to make.

Not a Little Thing

When we communicate the little things to the brokers, we open the door to earning their trust—and that's a big thing.

Take Away:

- Brokers fear losing control of the sale.

- You can mitigate this fear by communicating more often.

- Put more emphasis on building trust with the brokers than on just being likable. Remember, build trust and they will like you.

11 Remember to Become a Member

A lot of people want to know how to be successful. The secret is in the word itself: HOW - Help Others Win!
—*Unknown*

If you want to get on the inside of an organization, you should join that organization. It may be common sense, but very few builders in any given community even know that this opportunity is available. Fewer still would recognize the value if they did know. This is what I call opportunity. I tell the same thing to brokers. Many are interested in selling more new homes and working with builders, and my suggestion is simple: join and get involved both your local and national home builders associations. I have made many amazing contacts in both my local and the National Association of Home Builders. I attribute much of my success as a top producing broker in my city to this amazing organization, but again very few brokers are involved. This gives me the edge when it comes to working with builders and developers. It's almost so easy and so common sense that many real estate professionals miss this opportunity.

The Cost

What does it cost to join your local Association of Realtors? Obviously, the dollar amount will vary depending on your location, but for most it's just a few hundred dollars a year. Here are some of the common benefits of becoming an affiliate member of your local Association of Realtors. (The list comprises some of the benefits in my Local Realtors Association. Benefits may vary for each association.)

- **Free publicity and recognition.** All new members are recognized at a Realtors Rally and receive a framed certificate of membership. You may display your literature and marketing material at no charge at the Realtors Association events. Also, your contact information can be displayed on the local Realtors Association website.

- **Networking and special events.** Affiliate members get special discounts at the Realtors Networking events, Annual Charity Dinner, Auction, and Installation Banquet.
- **Advertising and sponsorship.** Most local Realtors Associations have a website and publication that you would be eligible to advertise in.
- **Member information.** All affiliate members are entitled to one free membership email list per year. This information right here should more than pay for your investment in this book. Remember, the brokers have the buyers! You are able to obtain hundreds or even thousands of accurate email addresses of every broker in your city. How many builders do you think are taking advantage of this? Having a grand opening? Bam! Every active broker in your city is notified in one email blast. (Note: If you do send out an email blast it is very important that you put the list in the Bcc–blind carbon copy. This will protect your list and it is common email courtesy. Please consult with a tech person if you have questions about this.)
- **Discounts.** If you want to throw your own broker appreciation party or new home training event you can rent their facilities at a discounted rate.
- **Government affairs.** One of the most powerful organizations in the country for the protection of housing rights is the National Association of Realtors. Your local organization is probably very involved in the issues that affect your business. Members are usually invited to political events held by the Realtors Association. This gives you access to local politicians and your support will further strengthen the political advocacy of the Realtors Association, which is usually in line with protecting and promoting homeownership.
- **Committee opportunities.** If you desire to become involved as a leader in your local organization, the opportunities for that are available. But you have to be a member!
- **Community service activities.** Your local Realtors Association is very involved in multiple charitable organizations and volunteer groups. Why not lend a helping

hand or sponsor an event? It has been said, "If you partake of a community, then you should partake in the community." If you are a home builder or builder's representative and you have become prosperous in this industry, it is usually because the local residents are investing in your product. It may be time to give back. The happiest and most prosperous people I know are givers. As my friend Jack used to say, "Generosity is an attribute that doesn't cost you anything." Oh, and for those who are not interested in altruistic benevolence, there is a very pragmatic reason to get involved in service: it is just plain good public relations.

- **Information.** Want to know how many homes are on the market? Want to know how many homes have sold in the month of June? Your local Realtors Association has that information. It may be possible to know if there is a glut of two-story houses on the market or if there is a shortage in a certain price range. Can you think of someone who might handsomely profit from such information? If you guessed a builder, you would be right! Again, how many builders are just shooting from the hip or following their gut? Why not get accurate housing and market statistics from your local Association? One more thing, if there is news or legislation affecting housing, the Realtors are usually the first to know. If you are a member, you will know too.

- **Education and training.** Some of the top trainers and educators in the real estate industry are giving classes for the Realtors Associations. Get access to the classes and discounts on the admission. You might even want to give your own class. I will be sharing tips about why and how to do this in later chapters.

Benefits Outweigh the Cost

If you are already a member of your local Realtors Association, see how you can get more involved. If you are not a member, it is worth it to contact them and ask about the opportunities for a non-real estate agent to become an affiliate member of the local organization. Again, benefits vary in different localities but usually the benefits far outweigh the cost of membership.

Take Away:

- Become an affiliate member of your local Realtors Association.

- Get on the inside and get involved. You will be one of the only builders in your area who is doing this and the payoff can be huge.

- Identify those brokers and broker agents most likely to market your homes.

Part Three:
Power Presentations

12 Company Story

*Live your life from the heart. Share from your heart. And your
story will touch and heal people's souls.* —**Melody Beattie**

I recently read in a sales book that presentations were the
number one way to generate new business, followed by signage,
referrals, and, finally, advertising. If this is true, why do most
builders spend the most money on advertising while spending
little, if any, time or money developing a presentation that can
powerfully convey the company story, competitive points of
difference, and a call to action? I believe many builders have
bought the lie that their homes sell themselves. In over 10
years of full time, front line new home sales, I have never had
anyone tell me, "Quint, please just show us how to fill out the
paperwork. We are ready to buy; the home has already done
the selling." Instead, when they enter a model home, many
prospects use the phrase, "We are just looking." Remember,
when you hear that they are just looking they are not just
looking. They are also listening. Who are you, and why should
they do business with you? Why should they choose your
company? The answer to the "Who are you?" question is your
company story. This can often be bigger than the components
or building techniques in your homes.

Why Do You Build?

What is the reason you wake up every day to build or sell
your homes? If I could summarize a great company story in
two words it would be authentic passion. The first aspect of
a Power Presentation is creating and communicating your
company story.

Develop Your Story

Here are some components you may want to consider when developing your company story:

- **Authenticity.** Shakespeare wrote, "To thy own self be true." It's ok to be you! Brokers and customers will appreciate the real you more than the fake someone else.
- **Core values.** The core values of the Unites States Air Force are: Integrity first, Service before self, and Excellence in all we do. I still remember these values even after being out of the Air Force for more than a decade. What are your core values, and do the members of your team know them?
- **Mission driven.** Nothing can stop a company or individual with a passionate pursuit.
- **Excellence.** Mediocrity is the disease of the day. Customers want to deal with people and companies of excellence.
- **Vision.** People want to know where you are going as a company.
- **Passion.** Facts may inform, but it is passion that persuades.
- **Altruism.** There are two kinds of people in this world: givers and takers. If buyers believe you are not a "giver," they will "take" their business elsewhere. Find small ways to let people know you care.
- **Community minded.** If you partake of a community, people want to know how you partake in the community. In other words, how does your company give back?
- **Stewardship**. Many home buyers these days care deeply about environmental protection. Help allay any concerns they may have by explaining the little things you do to be a steward of the environment.
- **Leadership.** Customers want to believe that the workers in your company are happily being led and not just managed by fear or coercion.
- **Accountability.** Nobody is perfect. Therefore, no company is perfect. Let your customers know that your company and employees are held accountable for quality and service.
- **Culture of quality.** This goes beyond the home itself and is demonstrated in the cleanliness of the jobsites and attention to the little details.

- **Personal development.** Where, or how, is your company making a difference in the lives of its workers? Maybe you helped pay for a worker's higher education or personal development. The bottom line is, if the customers feel that you don't care about your workers, how and why would you care about them after they purchase a home from you?
- **Develop a great team.** If you are more than a one-man show, you have a team. If you have a team, be proud of that team, and talk about what makes them special.
- **Communication.** In our world of constant connectedness, we are in many ways more disconnected than ever. Most people would rather speak to a real person instead of pressing "1" to leave a message. If you can convey to your customers that you will keep up good communication with them, and that they can contact you if they need help, you will have their attention and interest.
- **How do you define success? As the saying goes,** "Nobody on his or her death bed says, 'Gee, I wish I could sell or build one more house.'" Customers want to believe that the company they choose cares, and that you define success in ways that are deeper than just financial.

So, What's Your Story?

The above list is not comprehensive, but rather a list of ideas to spur interest and introspection as to what your company story could be.

One of the best company stories I have recently read was from David Weekley Homes. Go to davidweekleyhomes.com and click on the "Our Story" tab in the About Us section. The headline, "A Funny Thing Happened on the Way to Harvard..." almost forces you to keep reading to find out what happens. This is a great example of a key aspect of a company story—the importance of capturing the prospect's attention. After you read their "story," click on the "Executives" tab which takes you to a message from David himself. You can meet the executive team and the home team, then click the button that says, "Our Purpose." After spending some time on their site, I started to wonder if they were actually home builders or maybe some kind of community enrichment company. The point is that they effectively convey the message that their company has a higher purpose than just building a bunch of houses.

A great example of a video company story is for Harkins Builders. If you go to harkinsbuilders.com and click "About Harkins," a very impressive six-minute video plays, telling the company story. Not every company can invest in a six-minute company story video and not every company should. Sometimes less is more and simplicity can be the highest value. But, every company can and should spend the time and effort to find out what gets them up in the morning, what their vision is, and what they are committed to. A great company story should leave the audience or customers with the feeling that they want to root for you and your success. Remember, it doesn't have to be long. A company story can, and maybe should be, told in a minute or less. Start your story today.

Take Away:

- Create a company culture that makes buyers want to root for your success.
- The essence of your company story should be authentic passion.
- Make it your own. Be yourself and have fun developing your story.

13 Be Different!

Why fit in when you were born to stand out?
—Dr. Seuss

The second aspect of the Power Presentation is masterfully communicating the value differences you are offering. One of the biggest customer complaints I hear about home builders and their representatives is sameness. In other words, buyers want to know what distinguishes your offering from your competition. Therefore, when they say, "We are just looking," now you know what they are looking for. Buyers are looking for differences. In Chapter 12 Company Story, we covered letting the buyers know the Who and the Why of your company. In this chapter we are going to cover the What. Specifically, what makes your product or offering different and better? This is often referred to as Unique Selling Points (USPs), or simply, your differentials.

Three Aspects of Presenting Differences

There are three specific ways of being different that are important to remember. All three will help you pinpoint your own USPs, so keep reading.

Important to Whom?

The first point of difference ought to be common sense: **Be different in a way that is important to the buyer.** You can't do that if you don't listen and find out what is important to them. How do you do this? Simple—ask a question, such as, "Mr. and Mrs. Jones, please tell me, what is most important to you in your new home?" Notice I didn't say future home. Remember, we sell new homes. (That is a huge point of difference.) I like to say you should use the A.S.K. Technique. I created this acronym as a reminder to me to ask more questions and seek knowledge from my customers.

A–Always
S–Seeking
K–Knowledge

Be constantly and persistently curious with your buyers. When you are **Always Seeking Knowledge**, don't just hear with your ears. Really listen to what they are saying. Have you ever noticed that the word listen has the word *list* in it? That is a reminder that you are supposed to write down what is being said. As the saying goes, "The strongest memory is no match for the weakest pen." By the way, if you are taking notes, it is much more professional to use a high-quality, leather notebook rather than a sheet of paper from the printer. If you already have and know what your points of differences are, and the buying public agrees with you, then move ahead with creating your Power Presentation. If you are struggling to get an advantage over your competition, take heart. Often, builders don't really listen to what the buyers want. So if you do, you will gain the upper hand. Many builders are surrounded by hired people who will agree with anything the builder does, as long as their job is secure. In these competitive and changing times, it is important that we really seek knowledge and even criticism from our buyers and experts in the field. You will learn what is important and you can use this to create those points of difference that are important to your buyers.

Make Your Changes Smart

The second point of difference is where you put the knowledge you gained in point one into play: **Emphasize differences that are difficult for your competitors to reproduce or copy.** Look at these two conversations with a prospective buyer. Which difference gets his or her attention?

a. "Look! We use stainless appliances!" Customer: "Yawn!"
b. "Our garages are two feet wider and two feet deeper than the other guys'." Customer: "Hey, that's great! I need more storage!"

What makes the difference? A builder can easily switch out appliances, but it would be difficult to expand a garage. I've seen builders lose sales to home buyers who drive lengthy or oversized vehicles.

Another way of communicating a value difference that is difficult to reproduce is by emphasizing the intangible. The Japanese automakers were masters of this. In the 1980s, when the American car companies were boasting about double wishbone suspension and the torque of the engine, Toyota created the slogan, "Toyota, Oh What a Feeling!" Ok, so how do you copy that? Toyota's current slogan is, "Let's go places." The current slogan for McDonalds is, "I'm lovin' it." If their slogan was, "Great hamburgers at a low price," someone could, if they had the capacity, create a better quality hamburger at an even lower price. How do you copy, "I'm lovin' it?"

So, what are your intangible value differences? Are you a local builder? This can potentially be a big positive. Many buyers want to shop locally and support local businesses. Let buyers know you are of, by, and for the community—that you will be investing back into the community and that you care and will be there for them. Communicate that you have the nuances of the style and design in your local area, the stuff the big guys can't quite match. Maybe it's that magic feeling that your models give. Start today by thinking of value differences in your company and homes that are difficult for others to copy.

Superior Differences

The third point of difference goes hand in hand with the second: **Be different in a way that is clearly superior to the competition.** Saying, "Look, we use 2×4 exterior walls," would not be a superior point of difference. However, demonstrating that you use name brand, cast iron tubs versus plastic or steel tubs would be a great point to present, because cast iron tubs to most people are clearly superior to plastic or steel.

One important point to consider is the fact that if your competition doesn't point out their quality features, it is as though they don't exist. The building electrical code in our area mandated that every builder put a sensor on the side of the home that could tell if it was light or dark. When the sun went down the outside lights would go on, and in the morning sun, they would go off. I started to demonstrate and show this to the buyers and the brokers and they were amazed at the thoughtfulness and intelligence of my builder. It is quite possible that I was the only new home sales person in my city demonstrating this value point. Only once did I have someone say, "Hey, that is just code, isn't it?" I responded, "Not for used houses!"

Break from the pack!

What makes your homes and company different than the competition? The buyers want to know who you are, how you are different, and how those differences are superior to the competition. Because we live in the information age, many buyers, and even brokers, clump all of the builders together. It is your job to break from the pack of builders who are lost in the morass of indistinguishable sameness. Our primary jobs, as ambassadors and communicators of value to the broker and buyer community, are to be masters of creating contrast. A great title for our jobs would be Master "Contraster!"

Take Away:

- Remember, buyers want to know what makes you different

- The USPs in your product or company need to be important to your buyers, clearly superior, and difficult to reproduce

- Remember, if the competition doesn't point out their points of difference it is as though they don't exist.

- Don't forget to emphasize the intangibles.

- Remember—details sell. If you want to be terrific you have to be specific!

14 Position Your Company and Your Sales Team

*It is amazing how much you can accomplish when it doesn't matter who gets the credit. —**Harry S. Truman***

The third aspect of a Power Presentation is the *how*. First, notice there are four kinds of builders:

1. Builders who don't give any broker office presentations.
2. Builders who give presentations without talking about *who* they are.
3. Builders who give presentations without explaining *what* makes their homes and company different.
4. Builders who give presentations without explaining *how* the broker can get involved to sell their homes.

Sample Outline

Don't neglect the *how!* Here is an outline of an effective, five-minute broker sales presentation, giving a minute to each subject:

1. Establish rapport (recognize their company and individuals)
2. Builder's story
3. What makes your company differerent
4. How the brokers can get involved in selling your product
5. Potential rewards or benefits of selling New

The How is Simple

Now you are getting down to practicalities. Tell the brokers how to sell your homes!

1. What are your hours of operation?
2. How can they sell after hours? Are your homes on Lock Box?
3. Who is the main point of contact?
4. Do they have to do the paperwork, or is it in-house paperwork?
5. What is your website information?
6. When do they get paid and how much?
7. What title company do you use?
8. How can they register their buyers?
9. How long does it take to build a house?
10. Where are you building and what are the prices?

Again, this may seem elementary, but trust me on this. There are many builders who put on big parties for the brokers—giving away expensive prizes such as iPads and cash. However, when the party is over the brokers don't know who the builder really is, besides maybe the company name. Just last week I attended a grand opening for a new builder to the area. The balloons were up in the air, they had food and prizes, but the other brokers and I left with little or no real information about their product, company, or even whom to contact if we had qualified buyers. In my opinion, it was just another expensive party. After many of these events, the vast majority of brokers can't tell you more than a few details about the homes. Finally, and worst of all, they are left with no information as to how to sell your homes. Most brokers will not call you to ask, either.

A Major Reason Brokers Avoid New Home Sales

I believe one reason many brokers avoid new home sales is because they don't want to look silly in front of their buyers for not knowing how to sell them. This is easily avoided by outlining the simple steps they can take when they do have a buyer interested in looking at new homes.

Believe me, I am equally challenging and critical of the broker community for not making the effort to know more about the details of the builders in their area. But the fact remains that in most areas of the country, few brokers have any expertise when it comes to new home sales. This is unfortunate because, as I have stated many times in this book, the brokers have the buyers. So, until the brokers are inspired to learn about your company and your homes, it is your job to educate them when you are given the opportunity. What better opportunity than a grand opening?

Position Yourself for Success

One key to success in this area is to position your sales team as assistants to the broker community. It is important that you communicate to the broker community that they will not lose control over their buyers or the sale when they bring a prospect to your company. You and your sales team are there to serve the broker community and assist them in successfully selling more new homes. I have found this technique to be very helpful in creating positive and effective broker relations. During an office presentation, I have often asked the brokers, "Raise your hand if you have an assistant." Usually, very few brokers raise their hands. Then I make the statement, "You do now! I want you to think of me as your full-time, licensed new home sales assistant tasked with the assignment of helping you sell more new homes." It has been wisely said, "It's amazing how much you can accomplish if you do not care who gets the credit." Assist the brokers, give them your help and you will reap the rich reward of having a potentially unlimited pipeline of buyers.

Examine the Spirit of Your Approach

If we approach the broker community in a spirit of humility and service, they will respond in force with support for our companies and endorsement of our homes.

My acronym for the word S.A.L.E.S. is: "Service And Leadership Equals Success."

S-Service
A-And
L-Leadership
E-Equals
S-Success

Take Away:

- Never allow a broker party or grand opening to end without taking the time to let the broker know whom to contact and the steps for him or her to take if he or she has a qualified buyer.

- Position your sales team as assistants to the broker community. You and your sales team are there to serve the broker community and assist them in successfully selling more new homes.

15 New Beats Used

Believe in yourself! Have faith in your abilities! Without a humble but reasonable confidence in your own powers, you cannot be successful or happy. —**Norman Vincent Peale**

A successful business should know what its strengths and weaknesses are. In addition, you should know the strengths and weaknesses of your competition. One method of learning about your competition is through mystery shopping. There are many companies that offer services in this area, but it can be as simple as having a friend shop the competition to see what they are saying and doing. One thing I have noticed when shopping new home sales representatives is that they hardly ever tell you about the advantages of owning a new home. Oftentimes, they share the information or details about their homes, but almost never do I hear a representative ask, "Are you considering a used house? Please allow me to share with you the many benefits of owning a new home in place of purchasing a used house."

Not Just *One* Advantage

The fact that your homes are new is not just an advantage. It can be, as Jeffery Gitomer says in his book *The Sales Bible,* "A weapon of mass production." Almost invariably, the biggest competition to home builders is used houses. Many new home sales professionals struggle in selling against used houses. Home building is one of the only industries that struggles when it comes to selling against used homes. Can you imagine a sales representative from a computer store not knowing how to overcome this objection? A customer comes into a computer store and says, "I can get a five-year-old computer off of Craigslist for half the price of your new computer." The computer guy, who is probably not a natural salesman, would simply start chuckling, and say, "Why would you even consider buying a five-year-old computer?" You might be thinking, "But Quint, a home is *different;* real estate always goes up, right?" After 2007 we should all know that this is not always true, and in fact the argument could be made that the improvements to real property actually go down in value.

Be Prepared to Compare

If you were to compare the components of a used house to a new one, item by item, would someone really make the argument that the used house should be priced the same as a new home? Has the carpet gone up in value? What about the toilets or water heater? I like to ask customers who are considering used, "If you were in the market for a dishwasher, where would you go?" Nobody in the last ten years has ever said to me, "A flea market, Quint. I would buy one from a flea market." Let me be clear about something. I talk to many families who are struggling financially in this difficult economy. In truth, I am not against buying items from the flea market, yard sales, or even the Salvation Army. I would be upset, however, if I went to a yard sale and saw used sneakers for $60. Especially if I could buy the same ones brand new for $75. The point is, in many cases used makes sense if it is priced much less than the similar new item. What I am seeing in many markets, however, is that used houses are selling for the same, slightly less, and even in some cases more, than new homes.

Communicate the Advantages!

So why would anyone buy the used over the new? The answer is lack of education on the part of the broker community representing the buyer, and lack of skill and training on the part of the new home sales representative in his or her ability to contrast and communicate the advantages of a new home over a used one. Knowing how to effectively sell against used houses is a key skill to be mastered by any new home sales representative. Once this skill is mastered, one of the top focuses of your company should be to educate and influence the broker community. Your job is to sell the brokers on why they should be showing, selling, and promoting new homes to their buyers.

A Few Points to Start You Off

Here are some areas where new homes beat used houses:

- **Clean and fresh.** Never been lived in, ever! Or, died in!!!
- **State of the art.** Appliances, H.V.A.C. systems, and modern conveniences.
- **Customization.** Ability to personalize.
- **Light and bright.** Used houses can be dark and drab.
- **Energy-efficient and green.** The efficiency standards are much higher than even a few years ago.
- **Low maintenance.** Spend more time doing the things you love.
- **Safety.** New homes are typically more fire-safe and meet stricter electrical and child safety standards. In addition, they are free from mold, asbestos, and lead.
- **Engineering.** Modern technology allows builders to meet more exact specifications. They are not just guessing and building from a hunch.
- **Style and design.** Just as (apart from Halloween costumes) not many people wear bell-bottoms, there are many used houses that are simply out of style. In a new home you can get the newest and most desirable style and design.
- **Peace of mind.** New home buyers have the peace of mind that their homes are backed up by a written builder warranty.

Educate and Influence

Knowing the benefits of new homes is only one part of it. The real key is educating and influencing the buyers and brokers. This can be a tricky endeavor, because it is much more in the category of an art rather than a science. There are, however, some simple suggestions that may be helpful.

- **Be sensitive.** It should go without saying that if buyers simply can't afford your new home and a used house is all they can afford, don't bash used. Instead, encourage them and ask them to consider your offering in a few years if they are in a position to upgrade. This is also a good time to ask for referral business.

- **Use humor.** I carry several props in my model home. One is a urine detector. (Yes, this is a real item, and it can be purchased on Amazon.com.) I will never forget the first time I used this in my presentation. My buyers were qualified, willing, and able to purchase my new home, and I truly felt that the home I was offering was the best for their family. The husband, however, was boasting that there were a lot of homes out on the market that were priced less than ours. Because I knew my competitor's product, I knew he must have been referring to a used house. So I asked, "Are you referring to used houses?" His response was, "Well, yes. We are looking at several resale homes that are priced well below yours." I reached under my desk and pulled out the urine detector and handed it to the wife. Then I said, "I always recommend shopping and looking at everything, but if you are considering used, please borrow my urine detector. You may be surprised by what you will find in many used houses." The husband didn't know what to say and remained silent for a few moments. The wife looked down, then slowly turned to her husband, and quietly but firmly stated, "Honey, I am tired of looking at used houses." With a quizzical, but relieved, smile the husband said, "Ok, Quint, let's write it up."
- **Educate.** It is tough to argue with facts. If your windows are Low-E and the used house has standard "heat magnifiers," that is a material fact in your advantage. If the S.E.E.R. (Seasonal Energy Effiency Rating) in your home is 16, and in the used house it is only 10, let your buyers know. It may be worth it to ask what houses they are considering. Do some research and find the weaknesses and point them out. If you don't have the time or desire to see the home in person, Google the address and in many cases you will be able to search through pictures of the home on Trulia or Zillow.

I used this technique to sell against a house that one of my buyers was considering, instead of ours. This particular house was finished and the buyers didn't want to wait to have a home built with us. I viewed the home and realized that there was a sunken living room.

In the interview process I found out that the buyers' elderly mother was going to be living in the home, and the daughter was going to be her caretaker. I explained that the home they were considering would not be the best choice for them because of the falling hazard of the sunken living room. The daughter said, "Quint, you are right; I was really bothered by that and I have actually been thinking the same thing."

If the pictures of the used house they are looking at shows a huge tree in the front yard, let them know that it may be wise to double check that the roots are not going into the plumbing. The key is to only make legitimate observations based upon your expertise and level of product knowledge. You can balance finding fault with the home by pointing out actual items that could be considered positive. That way you are not just bashing the home, but offering valuable feedback. For example, you could point out that it looks like the owners have done a great job maintaining the yard. I have found that customers appreciate my feedback on the homes they are considering.

- **Use actual dates.** Every car company boasts about their newest released cars and, when they do, they add the date. The new 2017s just arrived and all of the 2016s are discounted to rock bottom prices. Why are builders not using this same approach in new home sales? This is something I have started to implement with great success. During a Realtors Rally or buyer presentation I say, "I am very excited about the early release of our new 2017 floor plans." If you are passionate and excited, the brokers and buyers will also get excited. Remember, facts inform, but passion persuades.
- **Change their perspective.** I don't believe that high-pressure approaches work very well when it comes to new home sales. The old saying holds true, "A man convinced against his will is of the same opinion still". In light of this truth, I created a new approach with buyers called, "High Perspective Selling^{SM}." Instead of trying to change the buyers, you will simply change their perspective. Here is an example when it comes to selling against used houses: Do you know what happens when people leave here and put an offer in on a used house? I usually get a phone call from a buyer who

calls excitedly, saying, "Quint, we finally got an offer on our home! All we have to do is make it through inspections and we will be ready to invest in a brand new home."

You could also say, "Did you know that one of the number one reasons people are selling their home is because they want to invest in and enjoy the many benefits of a new home?" I have had customers tell me, "Wow, I have never thought of that."

- **Be sensitive to the broker.** If a broker is showing several homes, including some of his or her own resale listings, you will encounter resistance and resentment if you bash used houses.

- **Utilize the right sort of prizes.** When I am doing a broker presentation, I often do a drawing for prizes. Along with Starbucks gift cards, I might give away a few urine detectors. I might say something like, "If you are showing used houses to your buyers, you can give the home a quick scan. You would be one of the only brokers in the world offering this service and courtesy to your buyers." People know I am the New Home Guy, and they will usually laugh and get that I am doing it in jest.

- **Germs.** If it is winter, I ask buyers if they have been looking at used houses. If they say yes, I will usually offer some hand sanitizer and say something like, "Please be careful; it is flu season, after all." It is important to let the buyers know you are just having fun and kidding around, but then get back to demonstrating the value of new.

- **Never lived in.** Simply pointing out to a customer that a new home has never been lived in or died in, showered or cooked in, and doesn't have any hair, toenails, termites or mold, can be very persuasive.

- **Death in the house.** There is a company out now called, DiedinHouse.com. They can run a scan and search through all police records to determine if anyone has died in a particular house. You may ask your buyers if they have done a search on their home.

Sales Rule Number One: Do No Harm!

I realize that with a few of the above tactics you may be thinking, "I could never say those things." If you don't feel comfortable with any of these approaches, the answer is simple and important—don't use them. Rule number one of sales and marketing is do no harm. I intentionally used some extreme examples of selling against used houses with the notion that it might spark some ideas about how your company and sales force can effectively highlight the value of new homes. Again, be sensitive to each situation, use humor, smile, and have fun. Especially if the buyers simply can't afford your new homes, don't bash used houses. The point is don't forget to highlight your biggest advantage and point of difference: your homes are **new**!

Take Away:

- Know how to sell against used houses.

- Be sensitive to your buyers, use humor, and, above all, do no harm

- When it comes to new homes, accentuate the positives.

16 Office Presentations

Be brave, even if you're not. Pretend to be. No one can tell the difference. —*Unknown*

Going to meetings is one of the most effective and under-utilized of all strategies when it comes to effectively working with brokers. Most brokerages have weekly or monthly meetings and your goal should be to present at these meetings at least once a year. Most builders miss this opportunity because they don't see the value, don't even know about the opportunity, or worse yet, are not welcome by the brokerage to present. If you have followed the suggestions in the earlier chapters, you have established a relationship with the qualifying broker, and should have good working relationships with the brokers.

Tips for Effectiveness

Here are some tips for hosting an effective office presentation for brokers:

1. **Schedule it in advance.** Sometimes openings to present are several weeks or months out, so don't wait until the last minute.
2. **Give a reason for the presentation.** Ask the QB for the privilege of taking a few minutes for giving a new home market update and making known the details of a new community or series of floor plans.
3. **Offer to bring food and host a breakfast.** This is usually very well received by the broker community and is a great opportunity to build goodwill.

4. **Find out about food allergies.** Talk to the QB, office manager, or receptionist about food preferences and possible allergies. If six agents in the office are gluten intolerant, offer an alternative. This shows respect, caring, and attention to detail. They will appreciate you for this.

5. **Be respectful of time.** Remember, they are not there to see you. They are held captive because of a mandatory weekly meeting. I have seen people abuse this opportunity, and it is awkward when they are asked to wrap things up by the managing broker or QB. Ask the person in charge how much time you have and then stick to it. Set a timer if you have to. When it goes off, just say, "Looks like I am out of time. I want to be respectful of your meeting, so please just contact me directly with any additional questions."

6. **Empathize.** When you are doing an office presentation, before you get right into your product and promotions, take a minute to empathize with the broker community. "We know that there are builders who don't understand the challenges brokers face in today's market, what with all the continuing education, E and O insurance, local fees, MLS fees, National Association of Realtors fees, and that's just to be open for business. To get just one qualified buyer, we at ABC Builders understand that it sometimes takes hours or even days of phone calls, prospecting and lead-generation. At ABC Builders we want you to know we understand and respect the work and sacrifice you make to be in this business and we are here to support you."

7. **Give a packet to each broker in attendance.** Make it easy for the brokers to find you and make sure they know where to go and whom to call when they have a buyer. Print a map, include floor plans, and add a printed builder story. These are just some ideas to stimulate your thinking in this area. The idea is to make sure you leave the brokers with something tangible they can use to increase their knowledge about your product and how to sell it.

8. **Remember the managers.** Be sure to express your appreciation and respect for the qualifying broker or the managing personnel who gave the opportunity to present. Maybe even present him or her with a gift or framed certificate of appreciation. Many companies put up certificates of appreciation from the local Boy Scouts and

Chamber of Commerce. Why not one from your building company?

9. **Make sure your presentation adds value and gets to the point.** After quickly establishing rapport and expressing appreciation, the quote by Saied Djavadi says it best, "Don't build me a clock. Just give me the time!" Give them three reasons why your community is a desirable choice for their buyers. Tell them 3–5 reasons why your product and offering is superior.

10. **Pass a basket and collect cards for a drawing.** This gives you the opportunity to get their contact information so you can follow up. I recommend drawings where everyone wins. One effective approach is to write a thank-you card and add a small gift such as a business card holder with your company logo on it. I encourage every agent to make separate business cards that say "New Home Buyer Specialist" on them. This is a great tip for brokers, because many buyers want to deal with a specialist, and many buyers want to look at new homes. You can put a note that says, "These business card holders are for your New Home Buyer Specialist cards. All the best, and happy selling!" You could even work with a local printer and offer a gift card for each agent to get business cards that say "New Home Buyer Specialist" along with their contact information. I got this idea from a lender who offered to buy agents their business cards as long as they put the lender's logo on the card. Genius! Many agents took the offer and were passing out literally thousands of cards with the lender's contact information and logo. It is important to remember that the gifts do not have to be expensive to be effective. One time I put a $2 bill in a small motivational quote book. After writing a quick thank-you card, I wrote, "Here's '2' your success in Real Estate!" The brokers loved it. I was able to effectively reach and establish rapport with hundreds of brokers on a relatively small budget.

Take Away:

- Be brave. Be brief. Be bold.

- Don't forget to gather the cards of those in attendance for follow up..

- Remember to include management and the receptionist on your thank you list.

17 Model Home Events

The Separation is in the Preparation.
– Russell Wilson

Model home events, such as grand openings and Parade of Homes, are a key component of a successful marketing plan. It is important that the brokers attend your company's model home event. I don't just measure success by attendance but on our ability to educate and communicate the value of our homes to the broker community. The question is how to have a well-attended and successful model home event while still being cost effective.

Tips for Success

Here are some quick tips for success.

*   **Invite the brokers.** This should go without saying, but many builders have public grand openings and don't even send an invitation to the brokers.
*   **Offer prizes.** A drawing for cash or a gift card is a great way to collect business cards for future follow-up and marketing. Offering prizes can also give more incentive to the brokers to view your new offering. My preference is to offer more prizes at less cost than to offer big prizes such as an iPad. This way more people win and it is generally better received. When doing a drawing for a prize, I also recommend making the price of entry a 5- or 10-question survey. In the survey, ask the brokers how your homes stack up against the competition. Brokers are professionals who see literally hundreds or thousands of homes every year and are on the front lines talking to the buyers every day. They can dramatically help your company.

- **Offer food.** Brokers are usually out and about, and if you offer lunch they will usually attend.
- **Don't compete.** Check to make sure there aren't any competing broker events scheduled for the same time as your event. It is a costly mistake to find out there is a mandatory broker course the same time as your grand opening. This actually happened to me. You can find this out by calling your local Realtors Association.
- **Get sponsors.** Banks, title companies, mortgage brokers, and even the developer, are great potential sponsors to synergize with. Many title companies, for instance, have budgets set aside to support local builders and brokers and can even help with taking care of some of the details, such as food. Get the people with whom you do business involved; you may be surprised at their willingness and ability to help you!
- **Take time to present your company and product.** I cannot tell you how many builder events I have attended where the builder never says anything about his or her company or the product. These are usually the most expensive ones, with bands and big prizes. At the end, everyone is fed and maybe even drunk, but nobody can recall the builder's offering or any points of difference. So the whole event was nothing more than an expensive party. Never forget the power of the presentation. You are there with a purpose, and your purpose is to contrast your company and product with the other builders in your marketplace. You built the model, paid for lunch, and offered the prizes. Take a couple of minutes to introduce or remind the brokers who you are and what makes your company different. This should not be a 20-minute speech. Just take a few minutes to thank the brokers for their support. Give a few points from your company story, what makes your product special, how to contact your company, and whom to contact in your company in order to bring their buyers to you. And, perhaps offer a brief model home presentation.
- **Invite local VIPs.** Real Estate professionals are, as a whole, very social and usually are very interested in networking. Personally invite some local VIPs. If they agree to attend, ask if you can let the local brokers know about some of the key people who will be attending your grand opening.

With their permission, you can add their names as VIPs who will be in attendance. Let the brokers know that your grand opening will have food, prizes, and networking. An example of someone you could invite as a VIP would be a sports figure, mayor, president of a company, or principal of a school. Also, consider the police and fire chief. You may be surprised at the willingness of VIPs to attend and support your event. Remember to foster and encourage networking at your events.

- **Invite organizations that can support your grand opening.** The chamber of commerce, the home builders associations, or any local trade organizations can be a great support to your grand opening.
- **Trade professionals.** You may also consider inviting the principals of the trade companies you do business with. I have sold many homes to the trade professionals who have helped to build the homes I represent. They believe in the product because they are the ones responsible for building it. Inviting them to your grand opening allows them to feel pride in their work when the local public or real estate brokers complement their work.
- **Five dollars for five minutes.** This is an adaptation of the presentation technique above. Instead of doing a giant grand opening and just hoping it doesn't rain or that it is not conflicting with the schedules of the brokers, you simply do several scheduled-by-appointment grand openings. This can be done office-by-office or even individual-by-individual. This works better if you have a sales person who is staffing the models during regular business hours. Here is how it works: Send out invitations to the brokers asking them to come when it is convenient for them during your set business hours, or to call to set an appointment. You make the offering good for a two week period or more. The point is, you spread the grand opening out over a period of time, instead of just trying to do the whole thing in a few hours. The incentives you give them are refreshments and a $5 gift card, usually to Starbucks, or just give them cash, and in turn you give them a five-minute presentation covering your key points of difference and getting them up to speed on what's happening in your area.

People may balk at this approach, but I can tell you, from my personal experience, it is effective. There are two reasons for this: First, brokers are generally interested and want to know about new homes, so it gives the brokers a reason to stop by if they are in your area. Second, it gives you the time to visit in smaller groups or one-on-one and effectively answer any questions they may have. This is difficult when you are in a room with 30 or 40 other people. Remember, the good brokers make money in this business by selling homes, not winning gift cards. The brokers you want coming to your events are the ones who are interested in learning about, and selling, your homes. I believe we too often get caught up in basically paying brokers through big prizes to attend our events. The prizes should just be an added convenience and incentive to attend, not the primary motivating factor for attendance.

- **Parade or Showcase of Homes.** Many builders offer drawings during these events. Few, however, offer broker-only prizes. Send an email invitation to brokers letting them know there is a broker-only drawing done at your model. If they only have time to go to a few homes on showcase, and most brokers in larger markets do not have time to view every home, they will be more likely to attend yours.

Take Away:

- Take time to present your company and product.

- Get sponsors and enlist the help and support of others

- Be creative and have fun.

- Don't forget to emphasize the intangibles.

- Remember the top brokers are there to learn about your product so they can sell your homes. Give them the information they need to be successful.

Educating the
Real Estate Community

18

If you think training is expensive, try ignorance.
– Unknown

One of the biggest challenges of working brokers is that they often have little or no knowledge of new construction. Why not create an educational event for the brokers about new home sales or about construction techniques? If your company can provide new home sales training to the brokers, it is like killing two birds with one stone. The new home sales training will have the effect of getting the brokers more confident and excited about selling new homes. And your affiliation and support of the event will reflect positively on your company and in the eyes of the broker community, which means the new homes they are most likely to sell are your new homes. To underscore the importance of this idea just check out part of this article written by David Fletcher with *Realty Times* from February 17, 2015.

"According to the National Association of Realtors, 67% of all new homes sold are sold to prospects introduced to the new home builder by a Realtor. Yet, according to a study commissioned by Builder Homesite Inc., 'nearly two-thirds of Realtors believe that builders are not offering useful training about how to sell new homes.'

At least home builders are offering training, which is more than can be said for Realtors. If Realtors were asked about their satisfaction with new homes training offered in their own office, an unquantifiable guess would be about 99% would say, 'We get no new homes training.'"

Tip 1: Offer Continuing Education (CE) Credits for Your Class

According to licensing law, real estate brokers are required to obtain 30 credit hours of education and training. Put together a one or a two hour introductory class on new home sales and construction and take it to the real estate commission for one time training credit. This takes a little bit of work, but the payoff can be huge. Because licensees have to get their credit hours every several years they will be very inclined to attend your class. Here is a sample outline of a two-hour new home sales class. You could take approximately 15 minutes for each of the topics below for a total of two hours:

1. The opportunities in new home sales.
2. Trends in new construction.
3. Energy saving building techniques.
4. Style and design.
5. Construction knowledge and components.
6. New technology of new homes.
7. Safety features of new homes.
8. Green building and sustainability.
9. Financing.

Tip 2: Use Topic Experts as Presenters

Not only do you not need to be the presenter, it might be better if you are not. When it comes to the style and design portion, find a local home designer and ask if he or she would be willing to volunteer some time. You can have the HVAC company talk for a bit about the new technology and energy savings of modern heating and cooling systems. I also recommend working with other local companies to sponsor the event. Title companies, lenders, trade professionals, and even your local Home Builder's Association may be willing to help cater food, donate door prizes, or help coordinate the event.

Trap 1: Making the Presentation Too Self-Promotional

The presentation should not be one big advertisement for your company or the trade that is presenting. This is an opportunity for you to give back, educate, and add valuable training to your local real estate community. You will be rewarded, but if you use this time as a self-promotional event they will see through your motives and be unimpressed. If you do have people or companies help or sponsor the event, you do need to make their contributions known to the group you are presenting to. If you do get sponsors, reward them by creating a poster, signs, or flyers that say, for example, "The food today has been graciously provided by ABC Company." A quick verbal announcement in front of the group of brokers will also suffice.

Trap 2: Offering a Class for Free

Sometimes offering your class for free is a mistake. It may be better if a small investment is required. This will add perceived value. Adding a beneficiary charity to be the recipient of all proceeds is a great way to protect the perceived value of the class without making everyone think you are doing this to make money. It also builds good will, even with the brokers who don't attend your class.

Example: New Home Sales and Construction Knowledge class – Approved for 2 CE credits, $10 investment – All proceeds go to the local Animal-Rescue clinic.

It Works

In a small southwestern town I was able to get a new home sales class approved for CE credit. We were able to bring in one of the nation's top trainers in new home sales. The class was approved for 5 CE credits and we charged $50 per person. This was a relative bargain for CE credit and the perceived value and subject interest was high, so the participation was great.

We had over 80 licensed real estate brokers attend this class. The proceeds covered all expenses and the whole real estate community was excited and revitalized about selling new homes. I believe no amount of traditional advertising would have had as positive a long-term result as what we accomplished with this training event.

Take Away:

- Brokers want and need training in new home sales.

  Don't make the educational event one big infomercial about your company.

- Enlist the help of others. Many top trainers are willing to support educational events.

- It's OK, and sometimes even a good idea, to charge for the class. Sometimes people don't value something when it's free

- If possible, get the class certified for broker CE credit.

Part Four:
Agreements, Commissions, and Bonuses

19 Clarity About Compensation

Nothing astonishes men so much as common-sense and plain dealing. —Ralph Waldo Emerson

There was a builder who was offering a slightly higher percentage of compensation than the other builders in our area. Upon further inspection, the reason was because the builder wasn't including the land or upgrades in the compensation. The commission was only paid on the price of the home, minus the land and minus the upgrades. The local brokers saw through the guise and were not impressed with this tactic. I am actually not against creative ways to compensate, as long as it is clear. If a broker sells a $400,000 home, but only gets paid on $275,000 he or she will be very irritated, especially if it wasn't explained clearly up front.

It is a mistake to advertise three percent commissions. It is better to say we pay three percent of the final purchase price of the home plus gross receipts tax. If someone could misinterpret your compensation plan, it should probably be revised.

Clarity, Always Clarity

Decide upon a compensation plan for your homes and put it in writing. Make sure the brokers understand the terms clearly. Ask them if they have questions. Make sure the qualifying brokers understand, because they are the ones who will take the side of the broker who feels he or she has been cheated.

Explaining Broker Commissions in Front of Buyers

The first order of business is to eliminate the word commission. Commission is a negative word. Professional Service Fee, or Fee for Service, is much better. Second, put the compensation agreement in writing. The time to explain the Professional Service Fee is not in front of the buyer when the broker is in your office. Take time to visit the offices, and pass out a broker compensation agreement that is clear and specific.

Don't Make the Mistake...

It is a mistake not to let the broker community know in advance and in writing what you agree to pay. If a broker has to ask about his or her compensation in front of his or her buyer when entering your model, you are not doing your job.

If you find yourself in this awkward situation, turn to the buyers and do the following: "We at ABC Builders do pay a Professional Service Fee to your representative. His or her services to you are free, the fee is paid out of our marketing budget." If the broker asks what the percentage is, tell him or her code 3. Code is just broker lingo for the percentage of the Fee for Service. If you offer a four percent Fee for Service, then it would be a code 4. This is only used in front of buyers, because when an actual number or percentage is used, the buyer may start to calculate how much the sales people are making, instead of looking and focusing on the home.

Once a broker came into my model and asked, "What's the commission?" Then she asked if there was a sales bonus—all in front of the buyers. Because I was trained and prepared in this area, I kept the sale going smoothly. Ninety-nine percent of brokers have much too much tact to ever put you in this awkward situation. But because I have had it happen to me, I must assume that I am not the only one. We should never rely on the buyer's broker to present and sell our homes unless he or she has been trained in this area. Selling and presenting the home is the job of the builder representative. The buyers' brokers are compensated for bringing us the buyers. Knowing how to smoothly overcome these kinds of awkward situations will help you to successfully move the sale forward.

Dealing with Broker's Demands

What if a broker demands a higher commission than what has been offered? Did you know it is against real estate law for brokers to say that there is a "standard" compensation or commission? Also, according to Article 16-16 of the code of ethics for the National Association of Realtors, a broker should not negotiate a commission as part of an offer. In other words, if you have a home and you are offering a three percent commission, the broker cannot say, "I will bring you a buyer if you offer me a five percent commission."

Standard of Practice 16-16:

"REALTORS®, acting as subagents or buyer/tenant representatives or brokers, shall not use the terms of an offer to purchase/lease to attempt to modify the listing broker's offer of compensation to subagents or buyer/tenant representatives or brokers nor make the submission of an executed offer to purchase/ lease contingent on the listing broker's agreement to modify the offer of compensation. (Amended 1/04)"

Don't Be Pressured

If a broker attempts to modify your offer of compensation, gently but firmly remind any broker of his or her obligations as a real estate professional under the code of ethics. If you are continually pressured, it would be wise to contact the QB and let him or her know about the situation. Brokers should never threaten you or ask to increase your stated commission or offer bonuses. That being said, it is also wise to make sure you are offering commensurate compensation with your competition. Again, there is no standard, but I have seen many builders who offer a three percent commission of the total purchase price of the home, and the brokers in many areas seem to be satisfied with this offering of compensation.

Take Away:

- Be clear about compensation.

  Let the broker community know in advance what your compensation is.

- Avoid the word commission, instead use, professional service fee.

- Know your rights and never allow unethical brokers to bully or pressure you to change your compensation offering.

- Make sure your compensation offering is commensurate with the competing builders.

20 Broker Bonuses

Do not hire a man who does your work for money, but him who does it for love of it. – Henry David Thoreau

My advice: don't offer broker bonuses. You may be thinking, with all the talk of appreciating and valuing brokers, that I would be in favor of bonuses, but I am not. The reason is simple: you already are giving the brokers a bonus! **Having the opportunity to sell a new home that is fairly priced is a bonus.** When a broker sells a used house he or she has to go through inspections, waivers, objections, and resolutions. This is time consuming, and oftentimes will jeopardize the whole deal if problems such as mold or termites are found. Again, if you are ever in front of a group of brokers and you have the opportunity, ask them to raise their hand if they have ever had a contract on a used house blow up after the buyers had the home inspected. You will most likely have the entire room raise their hands. Now ask them if they have ever lost a new home contract because of a home inspection. Invariably nobody will be raising his or her hand. I have never in my career lost a contract on a new home because of a home inspection. I am not saying there weren't issues. There are always items on a home inspector's list. A home inspector's job is to find problems. I just give the list to the builder and he does the paint touch-ups or the adjustments to the doors and then we go to closing.

Bonuses Can Send the Wrong Message

Understand this: the brokers need your product. Brokers should be grateful to the local builders in their area for taking the financial risk to provide quality new homes for their buyers. One potential problem with broker bonuses is that they can be interpreted as an apology to the broker community. "We are sorry for the lousy product and high prices. We understand that it will be difficult to sell, so we are increasing your compensation." Instead of giving bonuses, I suggest putting the money back into the home and add value and quality for the buyers. When a home is high quality and competitively priced, it is a pleasure for the brokers to offer and sell it to their customers. What a bonus!

Broker Bonus Promotions

I just recently saw an ad by one of my competitors: a national builder was offering six percent to the buyer's agent. What they don't realize is that by advertising the higher commission they could be shooting themselves in the foot. When the promotion is over, many of the brokers feel cheated when they only get three percent for selling the home that last week was paying six.

Also, the competitors can print that flyer up and use it against the builder in the future. How do they get the flyer? Simple. The brokers bring it over and say, "Hey, the other guys are offering six percent." A competitor can show the flyer to prospects who are trying to decide between the two builders and say, "I wonder why the other builder is struggling and having to offer huge bonuses to get his or her homes sold? Could there be something wrong with those homes or the area?" I am not suggesting that you use this approach. I just want to make the point that a savvy sales person can use your broker bonuses against you.

Bonuses Can Be a Trap

Here are some of the reasons I believe broker bonuses are a trap:

- It sends the message to the brokers that your homes are hard to sell. Again, bonuses can be interpreted as a price or product apology to the brokers.
- If a buyer finds out that his or her representative is getting

a bonus, he or she may feel that the broker is not working for his or her best interests.

- The majority of brokers are usually not persuaded by bonuses. Sometimes, it can actually make them feel guilty for taking them. Brokers are usually required to disclose the compensation with their buyers, or the buyers will ultimately see the bonus on the papers at closing and it can put the broker in an awkward situation. They sometimes have to explain that they didn't just sell the home to them to get a bonus. I personally know several brokers who always just give the bonus, if there is one, back to the buyer for this reason.

- Many builders who offer big broker bonuses don't have the best relationships with the broker community. This is for two reasons. First, when the bonuses are implemented, huge marketing dollars are invested communicating the bonus to the brokers. When the broker bonuses are over, it is almost never communicated. So, often a large number of the brokers believe they will be getting bonuses when in fact they are not. Second, builders who offer bonuses have to be very strict and judicious with giving the bonuses out because of huge financial burden. For example: I have seen builders who will drop the price to try to close the deal if a buyer comes in without a broker. He or she usually drops it to the exact amount of the broker bonus and commission. Then the buyers who do have a broker come back with their broker and now the builder has to stand by his or her policy of: "If your broker doesn't attend your first visit, he or she cannot be included on the deal." I find it much better to always have a set broker compensation that is not negotiated away. If a buyer comes back with a broker, I simply say, "I am so glad you are working with Tom, he is one of the best in the business! Let's make this home yours, shall we?"

- Almost every broker has experienced a loss of a commission when his or her buyers go directly to a builder. But when the commission is double what it is normally, he or she is apt to fight doubly hard to get it, or push harder to hurt the builder who didn't pay his or her commission.

- Bonuses, like buyer incentives, can become addictive. It is easy to start but often very difficult to stop. When builders go back to offering the normal commission the brokers who missed out on the bonuses feel slighted.

- It can change the motivation of the brokers to bring you buyers for money instead of the fact that your company and homes are exceptional. If the brokers were asked, "Why do you bring your buyers to Quint at ABC Builder?" the answer should be, "Because I know my buyers will be very satisfied and I believe in the company, the product, and the people working there." not, "Because of the sales bonus."

A Good Answer to a Probable Question

I was doing a presentation for a brokerage and, at the end of my presentation, one of the brokers asked what our bonus was for selling one of our homes. I knew what she was asking, but instead of saying that we don't offer bonuses, I explained that the benefits of selling a new home were the bonus. She continued to push the issue by asking why she should bring me her buyers when she could go to the competition and get three percent of the purchase price, plus a $5,000 bonus. I asked the group to raise their hands if they agreed that the market was tougher than in the past. The entire group raised their hands. Then I explained that the price of the land in our area had been reduced up to fifty percent and we were hiring the best and most talented craftsmen, and they are working at a discount. The builder I represented was working at about half the margin of a few years before. I added that I was working for a discount compared to what a typical listing agent would expect to get. Then I asked, "Do you know why we are doing this? The reason is that we are not waiting for the market to get better; we are getting better. We are improving style and design, energy efficiency, safety, and quality points of difference to give your buyers the very best opportunity on a new home at the very best price possible. That is really what our goal should be, right?" Silence, and brokers nodded their heads in agreement.

Take Away:

- Broker bonuses can be a trap.

- The majority of brokers will not be persuaded to bring you buyers just because of a bonus.

- Brokers are most interested in having great products for their buyers. They want an easy sale and the buyers to be taken care of.

- Invest in the product and service of your company. Having a great product to sell is the bonus.

21 Broker Must Be Present

Today is a gift, that is why they call it the present.
—Alice Morse Earle

The best and busiest brokers sometimes have the highest number of qualified buyers. The problem is that it is physically impossible for these top-producing brokers to be everywhere at the same time. Using language that is inclusive, not exclusive, will go a long way to earning trust. There was a great article written on this subject by Anne Ladewig with *Sales & Marketing Ideas* magazine in the January/February 2015 edition. In the article titled "Real Estate Agents: Your Key to Selling More New Homes," Ladewig interviewed Kimberly Mackey, an expert in the area of builder-broker synergy. Mackey suggests:

"Instead of using language that cuts the real estate agent out of the process, use words such as, 'we're going to partner with you', and 'we will make you look like a rock star!' When you lessen that fear, you will bridge that gap." You could also say something to the effect of, "Here at ABC Homes, we appreciate the support of our local real estate community."

Don't Be Exclusive

Many times, buyers go out looking on their own and then bring back their broker to get the deal done. Having a firm rule that disallows broker representation if he or she is not present on the first visit could be a mistake in the long run. This is especially a problem when the builder's representative negotiates the broker commission off the purchase price of the home and gives rock bottom prices before the buyers are really ready to buy, and then the buyer leaves and comes back with their broker representative. This can be avoided by having a fixed amount that is never negotiated away by the new home sales person, because it is set aside as a fixed part of the marketing budget. Having this perspective also helps answer the question as to why you won't lower the price for the buyers who say, "But we don't have a broker." You can respond with, "We pay a Professional Service Fee to the broker community out of our marketing budget when they bring us buyers. It wouldn't make sense to say, 'I didn't use the billboard to find your homes, so please lower the price of the home.' Brokers are only one part of the marketing budget for this company." Obviously, you want to use tact and discretion in these situations.

Sometimes sharing this perspective can hold the line when it comes to buyers wanting to negotiate a broker's commission off the price of the house.

What to Do If the Broker Calls after the Contract is Written

This will happen, so be prepared. Again, I don't recommend broker bonuses and one of the reasons is because it makes situations such as this so much more difficult, as the dollar amount is much higher. I have, in many instances, written a contract with no broker and then a broker calls the very next day asking, "Did Mr. and Mrs. Jones purchase a home yesterday?" This can be a very tricky situation.

Be prepared for these situations and consider having a smaller Professional Service Fee that can be offered. The amount can be less than your normal offering: fifty percent or even twenty-five percent of your Fee for Service can go a long way. As can a hand written card thanking him or her for his or her service and explaining that, while you won't pay the full commission, you do value him or her as a professional and wish to do business in the future.

In my experience, many times the broker is embarrassed that the buyer has purchased a home and not included him or her, and is happy to have the compensation. This is for situations where the broker hasn't in fact shown the buyers the house, but has a working relationship with his or her buyers.

Again, many buyers can't make a buying decision until they have seen the used houses. I believe the brokers do earn their commission by eliminating the competition. When a Realtor thinks he or she has been cut out of a deal completely, sometimes a small service fee goes a long way to keeping strong ties with the brokers.

Bad Buyers

I have found that, in some situations, the buyers use and abuse the relationship with their broker. The kinds of buyers have the broker drive them all over town for several weeks, looking at dozens of homes. The broker works hard getting them prequalified and then the buyer suddenly disappears off the broker's radar. He or she sneaks off and tries to purchase a home directly from a builder to save money. I have literally had a broker bring me a buyer in the morning, and then had that same buyer come back to my model several hours later, without his or her broker, wanting to purchase the home but demanding a discount for not using a broker. I believed these buyers were acting unethically and I chose not to do business with them. If I saw a buyer stealing money out of a broker's purse or wallet, I would not want to do business with that buyer and I would immediately inform the broker of what I saw. In the same way, I make it a point to inform the broker when this happens.

Be Flexible

It is a trap to not be flexible in your approach with brokers. I recommend having policies in place, and still keep in mind that some situations are different and need to be handled differently.

Take Away:

- Be inclusive in your policy toward brokers.

- Keep a set amount in your homes set aside as part of your marketing budget. This amount does not get negotiated away

- Be prepared for brokers who come in after the sale is made. Consider a policy where a smaller service fee is offered if the broker did help.

- Every deal is different so be flexible.

22 Listing Agreements

Your life works to the degree you keep your agreements.
– Werner Erhardt

Are brokers the best choice when it comes to representing and selling your product? The answer is: it depends. There is a lot of hesitation on the part of the builders when it comes to having a broker represent their homes—and with good reason. Before the downturn in 2007, many builders simply had in-house, unlicensed sales people staffing their models, and the homes sold. After 2008, many in-house unlicensed sales people who were not selling left their jobs in search of another product that was selling better. Many builders turned in desperation to the licensed broker community and many brokers eagerly listed the product and attempted to take over the sales for the builder. The results were lackluster in many cases, and in some instances even complete failures. Make no mistake about it: new home sales is a completely different sector of real estate, requiring a unique skill set in those who are tasked with marketing and sales. Many brokers, not realizing these differences, simply listed the homes and put them in the Multiple Listing Service (MLS), then sat back and waited for sales. The finished homes (if they were presentable and priced right) sold, but beyond that many brokers were practicing outside of their expertise and sales in many cases floundered.

A Product That Doesn't Exist

The essence of new home sales is selling homes that don't yet exist. This can cause problems for brokers who are not skilled or trained in selling and presenting a product off of plans. Therefore, I highly recommend first that you find someone who is skilled in the basics of selling, presenting, and building relationships with brokers and buyers. If the person described above happens to be a licensed broker, great. Just know that simply having a real estate license does not qualify a person to sell new homes.

Tips

Agree to *agree!* Without an agreement in writing problems will arise. Here are some suggestions when it comes to listing agreements with brokers:

1. **Create a flexible listing agreement.** Exclusive Right to Sell is by far the most popular for many brokers. However, other agreements exist that may suit your situation better if the broker agrees to participate. Exclusive Agency gives one broker the right to sell and market your product, but you can retain the ability to sell and not pay a commission. This is a tool I used in working with several builders who just needed help during certain days of the week. It also helped me because they sat there all week and I just sat the weekends. This was a win-win. The majority of the buyers came on the weekends, so I made the majority of the sales. They were happy because their weekends were freed up and I was happy because the majority of the buyers came in on the weekend. This was a unique situation with a small builder, but it worked well. This is a fluid business and every situation is different, so flexibility can often foster a win-win scenario.
2. **Set a rule that the broker is not to take buyers from your model to sell his or her own listings.** This is an area where many brokers have harmed builder clients as well as their own reputations. Builders pay literally hundreds of marketing dollars for each prospect who comes to their model home. Brokers representing builders need to understand and respect the buyers coming to the model as guests of the builder. If you allow them to take your buyers, then why not the couches and refrigerator? This is simply unacceptable.
3. **If a broker is representing your company,** he or she should not be permitted to sell the homes of competing builders. This can be stipulated in the listing agreement. Only under

very rare situations should there be an exception to this rule. An example is if a builder is out of home sites and is waiting for new land to be developed. Allowing your broker to sell other competing builder homes may keep the broker financially afloat while the new product or land is being developed. The key is that the builder and the broker have to be in agreement, and permission needs to be granted.

4. **A big part of a broker's business is listing homes to sell.** I believe in some instances it is okay to work with a broker who continues to list used houses. Having the additional signs around town can actually help the broker pick up buyers who can be converted into new home buyers. This should only be allowed in rare instances. I believe a good time to allow this would be in the case where a builder needs a broker to list his or her homes in the MLS and sit part time in his or her homes, but simply can't support the volume needed for a broker or sales person to make a living selling just the builder's homes. For a higher-volume builder, allowing the listing sales broker to continue to list and sell used houses could be a mistake, because it could take the broker away from selling the new homes. Some of today's builders are using an approach where they hire and train non-licensed in-house sales staff, but then list the homes with a licensed broker at a dramatically reduced service fee to get the exposure in the MLS. I have seen builders pay as low $500-per-home-sold to a broker just to get their homes in MLS. (This is not the service fee for selling the home, but rather just the administration fee and fee for listing in the MLS.) This can be a synergistic relationship, and many brokers are willing to do it because it inflates their sales numbers and makes them look phenomenally successful. It can work out very well for the builder because of the power of the broker community and the MLS. It gets his or her homes in front of the brokers, who have the buyers.

5. **I recommend that builders include a clause that allows them to cancel the agreement if the relationship with the broker doesn't work out.** This is not to be used so that you can avoid paying on homes under agreement, but rather, it gives you an out if you hire someone who simply doesn't have the skill set required to sell new homes. The same goes for the brokers. I recommend that builders allow them to

leave or break the listing agreement if they are not happy or feel unsuited to the task. The last thing you want is a disgruntled broker representing your homes.

Take Away:

- Be inclusive in your policy toward brokers.
- Keep a set amount in your homes set aside as part of your marketing budget. This amount does not get negotiated away
- Be prepared for brokers who come in after the sale is made. Consider a policy where a smaller service fee is offered if the broker did help.
- Every deal is different so be flexible.

Part Five:
Tips and Traps

23 Ask the Brokers

The fool doth think he is wise, but the wise man knows himself to be a fool. – Shakespeare

One of the biggest traps builders face is not getting advice from the people who know the buyers, market, and competition the best. Yes, the brokers. It can be as simple as calling a broker and asking for advice about the market, product, or future recommendations. You may also consider a round table lunch where you take three to five top brokers to lunch or coffee. There are two kinds of brokers you should interview. Those who have previously sold your product and those who have not. Most brokers I know pride themselves on being knowledgeable about the market and will be very willing to assist you, if you ask them in the spirit of humility and respect.

Questions to Ask

Here are some practical questions that you may ask your local brokers:

1. How is the market?
2. How does our product compare to our competitors?
3. How does our product compare with the resale homes on the market?
4. What are your buyers looking for most in their next home?
5. What weaknesses do you see in our company?
6. What ways could our company serve   our buyers better?
7. How can we get better at reaching out to and serving the broker community?
8. What do you see as our biggest strength in the market place?
9. Do have any recommendations for our marketing?
10. What are we missing? We know that you are on the front

lines and may have a different perspective and see things that we can't.

Humility Fosters Respect

The precise questions are not as important as the spirit behind the questions. If you ask your local brokers in a spirit of humility and learning, they will be honored by your questions. Simple things such as taking notes and going back to their answers to clarify their opinions goes a long way to building respect and rapport. It is a well-known truth that people are honored when they are asked their opinion. You may be surprised by the eagerness of many brokers to help your efforts to learn and improve. Some of the richest and most wealthy builders and developers I know are sponges when it comes to gathering information from the front lines.

The Other Side...

I have also seen builders and developers make mind-boggling mistakes when it came to product and positioning. One builder I worked with spent thousands of dollars with a top national consulting firm to put together a comprehensive market analysis. I immediately recognized the builder was constructing the wrong product at the wrong price for our market. The company was constructing luxury two stories in a community where most of the buyers were retirees. When I started making recommendations as to its product, the company referred back to the comprehensive report that it had paid for. They were kind enough to let me review the report and I immediately discovered the mistakes. First, it compared homes that were sold on the MLS to presale new construction. The problem with this was that the finished homes at that time fetched a higher value than the homes that people had to wait six months to build. Second, it compared the price per square foot of one-story homes built on the south side of the highway and used that price for two story homes on the north side of the highway. Oh, and one more thing: the consulting firm never even knew about two builders who were building two-story homes very close to its area for $60,000 less. The competing builders were never picked up because neither of these builders put their homes in the MLS, which is where the consulting firm got its information. After two years of getting beat up, the builder finally decided to leave the marketplace. My guess is that the builder's financial losses were in the millions.

Just One Fact Can Make a Huge Difference

Sometimes, one key piece of information can create a tipping point in your favor. Even more important may be learning about potential mistakes in the beginning stages. One of my mentors, Bill Mitchell, reminds me, "A person doesn't have to be better or more skilled than you to be able to help you. Think about Tiger Woods. He has a coach not because the coach is better than him at golf, but because the coach can see things that Tiger cannot." It is the same way in new homes sales. In this fast-changing and dynamic marketplace, the more eyes you have on the front lines, the better an advantage you will gain.

Take Away:

- The more eyes you have on the front lines, the better advantage you will gain.

- The brokers have a unique perspective and they are on the front lines. Learn from them.

- Don't just trust national consultants or data crunchers. There is no substitute for having eyes on the front lines.

- Keep the lines of communication open and invite key brokers to let you know about market changes or developments with the competition.

24 Builder Info Binders

Accurate information is a key part of motivation.
—Mary Ann Allison

Most offices have a broker who is on desk duty during the working hours. These positions give the brokers an opportunity to get leads from prospects who walk into the real estate office or call on the phone. Remember, the brokers usually dominate in the area of capturing new traffic, and many of these buyers are ready to buy! This again is due to their superior office locations, internet presence, and signage. When a prospect calls in and the broker on duty takes the lead, he or she usually goes straight to the MLS search engine to look for homes. This is where the problem arises. Most builders have the ability and willingness to build pre-sale homes and many buyers are willing to wait to get the right home, but, all too often, the buyers never get the opportunity to even see the new construction because many of the potential new homes are not listed in the broker search data base or MLS.

Broker Handbooks

One technique that I have implemented is to take a three-ring binder, fill it with essential information, label it "Broker Handbook New Home Sales Guide and Quick Move-In Homes," and give it to the front desk to be used by the agent on duty to assist him or her if buyers call in, asking about new homes. It is important to keep this binder updated at least every week or two, especially if information such as pricing has changed. Be sure to add a disclosure stating that prices, terms, and offerings are subject to change and direct them back to your website or representative.

What to Put In the Binder

Here are some key points you may want to include:

- Company story
- Quick move-in homes
- Key contacts for sales
- Clear and specific directions to the community
- Floor plans and pricing
- Website information
- Procedure: a step-by-step guide explaining the steps in the new home construction process
- Community plan maps

Reasons to Make a Binder

Here are a few reasons to try this technique:

- You are essentially positioning the brokers in key forward positions with the information they need to capture qualified buyers.
- Many brokers will appreciate the support and be grateful for the helpful information.
- During the long, slow hours, many brokers will peruse your handbook and you will gain top-of-mind awareness with them.

The Main Investment

The total cost for this package should not be more than $4–5 dollars. So the real investment is your time. Covering every real estate office can be difficult and time consuming. As the saying goes, twenty percent of the people do eighty percent of the business. Instead of trying to produce a handbook for every real estate office, why not just include the top five real estate offices in your area? If the results are positive you can always increase the circulation. The key is to keep it updated. Once a month, or ideally once a week, should be your goal. Again, include your website in your handbook and keep the quick move-in homes updated and accurate. You may ask, "Why not just skip the binder and focus on the website?" The binder shows that you care, and it is something tangible they can touch and hold. Brokers will appreciate your support and at the least will look through your binder when things are slow. This allows them to learn more about your company and homes. You will most likely be the only builder in the area doing this, so it is just one more way that your company can break from the pack of other builders who are not even attempting to reach out to the brokers. Also, this gives you a creative excuse to stop by and network with the brokers. When the other builders are dropping off candies, cookies and doughnuts you will be offering something of value that will help the brokers be more successful in selling your product.

Take Away:

- Many qualified buyers are captured by the many real estate offices in your area. Providing a builder info binder can give the agent on call the key information he or she needs to convert these prospects into buyers of your new homes.

- One of the most important items to include is quick-move-in homes and contact information for the brokers if they do have a buyer.

- For a small investment this may create big dividends.

- Remember, you don't have to have one in every office. Start small and grow this program, if results are positive.

If They Can't Find You, 25 You Don't Exist

Direction is so much more important than speed.
Many are going nowhere fast. —Unknown

At a local city council meeting, we were trying to get a new sign ordinance passed allowing the builders and developers to put directional signs on the streets leading to the community. The current sign code did not allow directional signs going to the community, and any directional signs that the builders put up were quickly confiscated by a codes enforcement officer and fines were even given out. The developer was the first to speak, and he gave a very detailed description of how the signs would look and what sizes we were requesting. The feeling I got from the discussion of the council members was that they seemed to feel that directional signs were unnecessary. When I was given the opportunity to talk, I introduced myself as the guy who watches dust and tumbleweeds go by my model home when customers can't find the community. I named the community where I was selling and I asked the city council to raise their hands if any one of them could give me directions on how to get to it. No one raised his or her hand. I continued to explain that neither could the potential new home buyers.

How Many Brokers Know Where You Are?

If you are a builder, I have news for you. Brokers as a whole do not know how to get to your community. Yes, this is a generalization, but I hope I have your attention. Don't believe me? Just call a local real estate office and ask to speak to the qualifying broker, or just any broker who is available. My experience is that very few know where the new home communities are in any given city. If everyone knows where to find your homes, you are the exception. If you are like the majority of home builders, please know that many top-producing brokers in your community would not be able to find your community if you offered them $1,000 cash. Let me illustrate this reality with the following story. I had been selling in my community for a little over six months. One day, two fire trucks pulled up and about eight firefighters jumped off the truck and started walking toward my model.

As I greeted them I noticed they had a confused look on their faces. One of the older firefighters blurted out, "We need to know about these new communities. We had no idea all these homes were being built out here." I got them all cold waters and joked that my job was to make sure the local buyers and brokers knew how to find the community, not fire fighters and law enforcement. After a few laughs, it dawned on me that if the fire fighters don't know this community exists, what about the local brokers and buyers?

Why People Can't Find You

Here are a few reasons why people can't find your community:

- Online maps such as Google probably haven't caught up with the development. This is true in the community in which I am currently selling. I have been selling there for over two years and, if you tried to find it on Google maps, you would get lost.
- Unless you have finished homes that are listed, they are not being featured on the MLS.
- Trulia and Zillow get their information off of the MLS.
- Most Realtors search the MLS when looking for homes for their buyers.
- Usually streetlights don't get installed until the community is complete, so oftentimes it is dark in the evening and it gets overlooked.

What You Can Do About It

Here are 10 quick-fix items to help you get the buyers and brokers to your community:

1. Make sure directional signs are available going to the community. If you need to, find a private residence and ask for permission to use their yard. It may be worth paying them a small fee to get their cooperation. (Check city codes and sign ordinances and be sure your signs don't block the views of the vehicles at intersections. Remember, safety first and check with legal counsel if you are concerned about liability issues.)
2. If you do pay for billboards, get the ones that are en route to your community and make sure it has a big fat arrow pointing in the direction of the community. Remember, less is more. Leave the selling up to the sales people; use the signs to get the customer to the community.
3. Put directions on the promotional material.
4. Even better, include a map. If Google maps does have your community updated in its system, use its maps with graphic designs outlining the route to the model.
5. Put the GPS coordinates on the promotional materials.
6. Have detailed directions to your community saved to your computer as well as your phone. That way, if you have a buyer or broker asking for directions, you can send them a quick text or email. This is something I learned after about five hundred times of verbally trying to explain directions and typing directions out each time someone asked.
7. Sometimes the city is behind in getting street signs up. Make phone calls, send emails, do whatever it takes to get those street signs up. It is very frustrating to buyers to be told, "Turn left after the big pile of dirt."
8. Put exact mileage into the directions. If you are going through the effort to write out directions, why not be specific? "Go 0.9 of a mile, then make a left on Emerson Street," is better than, "Make a left on Emerson." Ever been driving behind people who are obviously looking for the right street? They slow down each time they pass the next street sign so they can read it. Sometimes the street signs are hard to see, but nearly everyone can read his or her odometer.
9. Put a Quick-Response (QR) code on your marketing material that says, "Scan here for directions."
10. Make a short video giving directions and showing the different landmarks as you get to the community. This can

be added to an email and is effective in getting the brokers and buyers out to your community.

First Experiences Count

By helping buyers and brokers easily find your community, you make their first experience a good one. Nothing is worse than trying to establish rapport with buyers who just spent 30 minutes trying to find your community. And brokers like to be in control. Being lost is pretty much the epitome of being out of control. Do everything you can to make sure your community is easy to find and you will not be left behind.

Take Away:

- Many brokers do not know how to find your community
- Taking a little time to make sure the brokers and buyers are able to easily find your community will pay big dividends.

26 Flags, Signs, and Promotional Materials

The early bird gets the worm, but the second mouse gets the cheese. —**Unknown**

One trend that I have noticed since the downturn of the market in 2007 is multiple builders in a community. Whereas in the past there were one or two builders in a community, now in some areas we are seeing five, ten, or even more in a single new home community. Whether this is happening in your locality or not, the fact remains that all too often builders get lost in the sea of competition and can get missed by potential buyers and the local brokers. This is the equivalent to participating in a relay race and dropping the baton.

Catch the Traffic!

Here are some tips for successfully capturing the traffic that does come to your community:

- Don't build your model on the corner, but rather one in from the corner.
- Make the corner lot into a parking lot. Many builders are tempted to put their model on the corner, but this has three negatives. First, the customers who see the model home on the corner will want their home built on the corner as well. There are a limited number of corner home-sites, so this is a problem. Second, it can create a parking problem. Third if you decide to build a new model, the corner model will essentially be blocking it.
- Convert the garage into a sales center. This makes your model home look like a model home; it is less likely to get missed.
- Put a banner over the sales center entry. This works and it is also great for branding.

- Put flags around the parking lot and the model home. Flags move and movement draws attention. I have had my competition tease me that my model home looks like the entrance to the League of Nations building. If this happens to you, it means you are doing it right!
- If covenants allow, put spotlights on the ground shining on the model. Many buyers and brokers drive around at night. If your model home is bright, you will be in their sight.
- If covenants allow, put floodlights on the corner of the model home.
- Use reflective signs or name riders. Make sure this is legal in your area. Reflective signs can really grab the attention of the brokers and buyers after dark.
- Put up an "Open" sign. Just as important is to take it down when the office is closed. I have made this mistake and buyers don't like it. "The sign says you are open, but the model is locked."
- Put hours of operation and an afterhours contact phone number. Make sure this information can be read from the street.
- Put a lock box on the front door so brokers can show the home after hours.
- When you leave the model at night, leave on the front and back exterior lights, any floodlights, and a hallway light. This is also important for the homes under construction, if they have electricity. When my company did this, besides providing cheap security, the area where we were building just looked much brighter during the night. People could see the homes and it produced a feeling of comfort and momentum for my builder. This is in stark contrast to one of my competitors who insisted on blacking out the model and homes under construction. The extra light is essential at night when many buyers are driving around checking out communities. I am astonished to see many of my competitors' models completely blacked out.
- Put signs up on the home sites that you have available or optioned. Many builders and sales people don't take the time to do this but it really pays off. A good many of my contacts come from people who see my sign on a home site that they are interested in.

Don't Be Shy

When brokers get to your community, let them know you are open for business. I have had several people comment that I had too many signs around my model, but nobody has ever told me that he or she couldn't find my model home once he or she was in the community. This is not true of my competitors. Remember this quote from the famous Zig Ziglar: "Timid sales people have skinny kids." If buyers and brokers are in your community, make sure it looks like you are open for business.

Take Away:

- Just because buyers and brokers make it to your community doesn't necessarily mean you are out of the woods. With increased competition you have to break from the pack and capture their attention and interest.

- Do the little things to make your model home shine.

- Don't be shy or bashful when it comes to signs, flags, and lighting. Remember, sometimes less is more. But when it comes to signage more often **more** is more.

27 Internet Marketing, Websites, and Social Media

*The problem with quotes on the internet is that you can never know if they are genuine. —**Abraham Lincoln, 1865***

It is beyond the scope of this book to completely cover internet marketing, social media, and websites. There are, however, several key areas of importance when it comes to effectively working with brokers.

1. **Respect confidentiality.** There is a social media trend to advertise the names and addresses of the buyers when they close on their home. Like this post: "Congratulations to the Sanchez family. They just closed on their brand new 2,500 sq. ft. home in the Willows." This kind of post is a problem because it breaks confidentiality and possibly fair housing laws. Brokers are held to a strict code of ethics, one of which is confidentiality. In today's complex world there can be many reasons people may want to keep their buying decisions confidential.

2. **Follow fair housing laws.** Because certain names may give away the race of your buyers, it is not wise to use buyers' names. Even saying whether it is a family or a single person can cause problems with fair housing laws. One broker I know got into trouble because he advertised that a home was within walking distance to the store. This was discriminatory against handicapped people who can't walk. The bottom line is you need know the laws, follow the rules, and seek guidance.

3. **Do no harm.** Rule number one of marketing, whether online or through physical mediums, is do no harm. I have seen many builders advertise deep discounts over social media, only to have the buyers currently under agreement come back to the negotiating table and demand the same discounts and incentives. Many times a builder under a deadline to move five already-built-homes highly promotes

his or her discounts and incentives. Then, after those homes are sold, the builder is confused as to why the new buyers refuse to purchase at the higher price. Remember that everything done online should be cautious and done with the attitude of do no harm.

4. **Make sure social networking doesn't turn into social "not working."** This is from Tom Richey's book, Probes. Obviously, every tool has its advantages and limitations. The point is to make sure your company is doing it right. After the downturn in 2007, many builders turned to social media to try to revive their sales. Remember, when airplanes were first invented, there were a lot of crashes. A similar phenomenon has occurred in this new field. Many builders using social media incorrectly have hurt, instead of helped, their cause. There are many qualified consultants and trainers when it comes to social media. The National Association of Home Builders and the National Association of Realtors can be great resources in this area.

5. **Give the brokers the information they need.** Brokers first want to know what homes are available for quick move-in. Let's be clear: if a broker is driving a customer around, he or she is usually in the market for a home fast. That is not to say the buyers wouldn't wait for a home, but if one was finished, most brokers would love for their buyers to purchase it. On my website, I have a button that simply says, "Quick Move-in." It has been very helpful for my broker friends to quickly print this off and show their customers. Next, I recommend that somewhere on your website you include the covenants for the communities. This is a timesaver and a big help to the brokers and buyers. Also, don't forget to give directions. Remember, if you are in a new community, there is a good chance that the brokers will not know exactly how to get there. And, no, they don't want to call you in front of their buyers to ask for directions. Why put them in that situation? Give specific directions and make them easily printable so they can quickly get to the business of showing and selling your homes.

6. **Give your company story,** but not your unique selling proposition. Remember it is not just potential buyers and brokers who are reading your website, it's your competition. Your company story is yours and nobody can copy it. It

is a big part of the essence and brand of your company. But if you advertise your top 20 points of difference, your competitors can simply copy and paste that, and now essentially their company is equal with yours. Instead, allow your sales professionals to do the selling of specifics.

7. **Explain the procedures for the brokers.** It is helpful if you have a tab on your website that says, "Brokers" or "Realtors." Under this is information for brokers when it comes to cooperation and steps to register their buyers with your sales professionals. A statement from your company expressing your gratitude and appreciation of the broker community can also be helpful. Again, if you are one of the only builders in your area doing this, your website will stand out to the broker community, and they will respond with gratitude and support for your company.

8. **Get creative.** For several hundred dollars, this or that company will guarantee that you come up on page one of Google. I realized that the whole purpose of being on page one of Google was to get people to check out my website. I decided that instead of paying these people to put my website on page one so that I may have people check out my site, why not just pay the people I want to check out the site? I offered monthly prizes to the broker community for simply checking out my website and registering under the tab "monthly prizes." I also directed my buyers to do the same thing. People started checking out my website, and its ranking grew organically. Now my website on a Google search is number one on the first page for new homes in my area. Be different, be creative, and if everyone is going left why not go right? It may be wise for your company to pay for search engine optimization and maybe even paid internet advertising. I just want to encourage you to keep thinking outside the box.

9. **Get sponsors.** "Okay, Quint," you might say, "I like your monthly prize thing, but that is going to get expensive, right?" Wrong! Now that my website is ranked higher, it puts me in a position to get others to promote their companies and offering. How can they do that? Simple—they sponsor the monthly prize. This month's prize is a free round of golf sponsored by the local golf resort. Next month's prize is sponsored by the local mortgage company. I have found

that many companies, and even subcontractors, are excited and happy about donating a small prize or gift card. It's win-win-win.

10. **Own and control the URL or domain.** A very successful builder friend of mine was moving into a new market and had renamed his company to give it a fresh new brand. I realized that he didn't yet have a website set up for the new company and explained that it was necessary that he got one. My recommendation to him was simple. Get the domain that matched the name of his company. The problem was, another person owned it and that person was asking $800 for it. I explained that it was a bargain because he would need that once things got up and rolling. He took my advice, but about a year later he told me that he didn't care for his website developer and he wanted to know how he could change companies. I thought, "Simple, you just fire them and get a new company." There was, however, one slight hitch. The web designer/developer had purchased the domain and had ownership and control of it. He could probably take the designer to court to regain ownership and control over the domain but it could be costly and time consuming. I thought this might just be a bad situation that happened to my friend, but it is actually very common. The bottom line is, if possible, own and control the domain or URL. Again, there are specialists and attorneys who can advise to the specifics of domain law. Every situation is different, so please consult with professionals who specialize in this area.

Take Away:

- Internet marketing, social media, and websites are not just the future, they are the now and we need to embrace these new and exciting mediums.

- It's important that we do it right when it comes to marketing over the internet. Don't be afraid to hire consultants who can help.

- Be prepared for brokers who come in after the sale is made.

  Consider a policy where a smaller service fee is offered if the broker did help.

- Every deal is different so be flexible.

28 Lean, Clean, and Mean

If you want to know the reputation of a builder, look at the model home. If you want to know the character of a builder, just look at the homes under construction. —**Quint Lears**

Most buyers do not understand all of the systems of a home, but they do understand what trash looks like. Many buyers are confused by the incentives that builders offer, but none are confused by what it means when there is food and trash in a home under construction. Few buyers really understand what S.E.E.R., Low-E, and H.E.R.S. ratings are, but they understand what a dirty home means. The message is clear, and it is understood that the builder has little respect for the homes it is building.

"Come on, Quint," you say, "buyers understand that a home under construction has to have some debris; it is a work in progress." I have seen builders desperate for sales offering huge financial discounts on homes that have chewed up sunflower seeds that had been spit all over the floors and counters. Food, trash, soda cans, and even scraps of wood are not acceptable. Many will say, "But what if the home is not complete?" My response is, "Do you want your homes to sell before, or after, they are complete?"

Most builders I know want their homes sold as soon as possible. If your homes are not clean and swept, it makes it very difficult for the marketing people to use updated pictures of your home. Because of this, they will often just use a rendering of the home until the home is clean and complete. This eliminates the possibility of putting up pictures of a home that are embarrassing due to trash and debris. When the home is complete, they take interior and exterior photos. By waiting until the home is complete instead of getting updated photos throughout construction, you are putting your company at a disadvantage. Many brokers and buyers who view the photos online will assume that your home is not anywhere close to completion and will eliminate your home because they are looking for homes that are quick move-in. Additionally, many of the real estate search engine portals will not pick up your home if you don't have a minimum number of photos. One of the easiest and most powerful ways to get a competitive advantage over your competition is to make sure the job site is clean. I am not talking about a hot mop. What I am saying is that a home should be swept clean and it should stay that way throughout the building process.

Tip: Hire Someone to Sweep the Homes Clean

Make sure your sales people are trained to point out the cleanliness of your homes under construction to their buyers. You will stand out in stark contrast to the builders who don't care to make an effort in this area.

When you are doing your broker event, make sure you don't just show your finished homes. Take them to a home under construction. Point out that it may not seem like a big deal, but it is important for your company to keep the homes swept clean while they are under construction. Nearly every time I say that to a buyer, the response is: "Actually, that is a big deal for us."

Offer incentives to the contractors for maintaining a clean and safe work site. Get a rolling magnetic nail picker upper. I like to walk around my community with this contraption. I also go to the homes just after the buyers move in and do a quick check around the exterior of their home for nails. This is a great opportunity to thank the buyers again for their business and ask for any referrals. I had a sales associate make the comment that I shouldn't use my magnetic nail pickup roller because it made the buyers focus on the negatives of new construction. This may be true, but I have found that most buyers really appreciate the fact that I am doing something about getting the community safer and cleaner. Honestly, I am a little bit selfish in my motivation, because after being in new home sales full time for over 10 years, I can tell you I have had my fair share of flat tires.

Trap: The High Cost of Low Cleaning Standards

This is what the trash and debris communicate to the buyers:

- **Soda cans.** This home will have an infestation of ants.
- **Food.** The home will have ants, and maybe there is food still in the walls.
- **Beer cans.** The workers are drunk and are going to make mistakes and do sloppy work.
- **Wood scraps.** It's a safety hazard to workers and potential buyers. Also, there may be the perception that the workers are wasting materials.
- **Particulates.** Did they sweep when they put down the carpet? I think I feel debris.
- **Nails.** This is unsafe for my children, and the builder is careless.

Take Away:

- It really does matter that the homes are kept swept clean.
- Use positive incentives to get the trades to participate in keeping the homes clean.
- Consider hiring a helper who can sweep the homes clean.
- There is a high cost to low cleaning standards. Make cleaning the homes under construction a top priority for your company and you will see positive results.

29 Putting it All Together

Never, never, never give up.
—Winston Churchill

Many builders don't realize the power of cooperating with the brokers and will not use the principles outlined in this book. Others will scoff at the idea of increasing broker outreach and participation. Because of this attitude in other builders, if you adopt the principles and techniques in this book, you will increase new home sales and gain market share. Once you have gained synergistic broker relationships, it is important that you maintain the momentum. Real estate is always changing and it is important that we keep up with the trends. Relationships of any kind take time and work and they are not always going to be easy. We cannot rest on our laurels. The fact that you have read this far tells me that you are a student of sales and are interested in mastering the subject. Remember what Arnold Palmer said, "The road to the top is always under construction." Never stop learning and never quit.

Key Points to Remember

Here are some key summary points to remember:

1. Foster a company culture that is positive toward the broker community.
2. Build relationships with the QBs and office staff at the brokerages.
3. Make the brokers your customers and do little things to show gratitude and recognition.
4. Create and communicate a positive and authentic company story. Master your presentation to the brokers.
5. Know how to tactfully and effectively sell against your biggest competitor: used houses.

6. Learn from the brokers. Remember, they have frontline knowledge and market wisdom.
7. Keep the homes under construction swept clean.
8. Continue to leverage websites, social media, and internet marketing.
9. Make sure the directions to your model are clear and included in your marketing. Remember, if they can't find you, you don't exist.
10. Effectively communicate to the brokers how to sell your homes when they have a buyer.

30 Bonus Section: Community Suggestions and Action Points

The difference between ordinary and extraordinary is that little extra. —*Jimmy Johnson*

The purpose of this bonus section is to give some quick tips and action points to help you avoid costly mistakes in the early stages of a community. I felt that this material, although not directly related to builder-broker relationships, was important to include. I have witnessed numerous communities suffering from lackluster sales because of mistakes that could have been easily fixed if caught early.

Names

Consider a community name carefully and, if possible, own the domain. Also, consider and carefully choose the names of the streets. I am baffled by some of the names that developers have chosen for the streets in their communities. Would you rather live on Turkey Knob or Mountain View? Also, consider that people living in your community will have to give their address to friends and family. If possible, make the spelling easy and avoid words that would be confusing. If you were a home owner giving out your address, it is not likely that you would have to repeat how to spell Oak Tree Court. However, if you choose names such as Vista de Nunatak, you are not making it easy for your future home owners. Lastly, if possible, avoid street numbers that could be offensive to your buyers. In the community I am currently selling in, every street has one address that ends in 666. Guess which ones I am struggling to sell. Why makes things more difficult than they have to be?

Signs

The signs in a community should advertise and highlight the community, not just the real estate brokerage marketing the community. By real estate law, if a broker is marketing your community he or she has to put the name and number of his or her brokerage on the signage and marketing materials. But, that doesn't mean the brokerage franchise information has to be the biggest thing on the sign, or that you have to use the same standard real estate brokerage signs that are used to market used houses. I suggest making custom signs for your community that support the brand and are consistent with the theme of your community.

Keep it Clean

As far as it is possible, fix anything blighted in and around your community. Many builders are willing to spend many thousands of dollars on paper or internet advertising but don't want to invest in keeping a community clean. If a neighbor isn't taking care of his or her yard and future buyers have to drive past that house to get to the new home sites available for sale, consider helping that home owner clean his or her yard. It may be that he or she is old or physically incapable of maintaining his or her property. Consider it a community service project. Let's say the people are just lazy and don't care about their yard. The fact remains that it will negatively affect your bottom line. Call the city and codes enforcement, organize a neighborhood cleanup, and hold sloppy trade professionals accountable. Bottom line: keep your neighborhood clean and remedy blighted areas or properties.

Weeds Kill Urgency

This is the same principle as keeping a jobsite clean. You are selling at every stage, from the moment the community area is selected, through construction, until the last site is sold. Keeping everything looking as good as possible from start to finish helps create and maintain the look of the neighborhood.

The higher the weeds in your community, the lower the urgency to buy will be.

Identify Home Sites

Put small signs that identify the home sites in your community. Also, put stakes that clearly mark corners of the home sites. As the saying goes, "Walk the home site and you will write." People are much more likely to sign their name to the deal and purchase the home if they walk the land where it will be built.

Releasing Home Sites and Incremental Development

Give the best effort to build out sequentially to avoid leaving people in a construction zone. Have a plan and order of operations. Consider releasing home sites in phases. This eliminates the feeling of the home sites being picked over and also creates more urgency. Ideally, you want to start in the back of the community and work to the entrance. Many communities sell a bunch of homes near the entrance and then the models and homes for sale are blocked from view.

Model Home Location

As a rule of thumb, avoid putting the model homes on an oversized or corner home site. A model home should be on the most average and standard-size home site. The purpose of the model home is to sell pre-sales or homes that don't yet exist. When the model home is on the corner then every buyer wants a corner. It's hard to tell a customer, "Yours will be just like the model home but on a smaller home site and without a view." If the model home is on an average home site it is easy to sell up, "Your home will be just like the model but you will have a premium home site that is larger and has a better view."

Sell the Differences in Time and Space

Instead of telling prospective customers that the home is only 15 minutes from town, you could say, "Driving an extra six minutes can get you four times the land." It turns a possible negative into a positive.

Consider the Words You Use in Your Ads

Instead of marketing 0.5-acre home sites you could say, "The average sized home site is 21,780 sq. ft." Sometimes changing the wording can get the customers thinking and taking notice of your offering.

Pricing

Everyone wants a premium home site: no one wants a lot premium. After a few years many communities are ninety percent sold out, with the last ten percent of the blighted home sites remaining, thus creating an unfinished feeling within the community. In the end, these home sites are usually sold with steep discounts. I recommend pricing any blighted home sites cheap to sell fast, and get them sold first. Also, start low and raise the prices slowly as the community develops.

Sell the Big Things First

If someone isn't sold on your part of town, there is no sense in trying to sell him or her on the granite counters. First sell him or her on the city, your part of town, community, home sites, and finally the floor plan and amenities in the home.

Get it done

Sometimes builders will allow a home owner to do their own front yard landscaping or even exterior finishes at a future date after closing. I even saw one builder close a home that didn't have a driveway. Sometimes the buyer is a contractor or even a landscaper and they want to save money and be able to add their personal touches to the unfinished part of the home after closing. The problem is, with all of the pressures of life, that later date never comes and the yard and property often times becomes an eyesore. Don't let this happen to your community. Create a standard of practice and stick to it.